Reasonable Books = Reasonable Prices

Irina Kleyman

Exceptional for Home Schooling

Math Help for 2-Year College

AF818403

First Edition

Reasonable Books = Reasonable Prices

Irina Kleyman

Exceptional for Home Schooling

Math Help for 2-Year College

I dedicate this book to my children and grandchildren.

Acknowledgments
Special thanks to Michael G LaFratta, Vladimir Kleyman, who has assisted in bringing all 10 books to market.

Author's note: In the many years that I've taught mathematics, I have been informed by many of my students that my methods for teaching have helped them tremendously in learning math. I have written these books in order to continue to help others learn math effectively.

Math Help Grade 1
Math Help Grade 2
Math Help Grade 3
Math Help Grade 4
Math Help Grade 5
Math Help Grade 6
Math Help Grade 7
Math Help Grade 8
Math Help for 2-Year College
Math Help for University (Basic, Algebra, Geometry)

ISBN: 978-0-578-85650-6 Paperback

Syllabus Math Help for College
Syllabus includes topics, what students need to know

#	Summary (Sum)	Topics Page - p	Pages 1	Pages 2
	Chapter(Ch)	Preface	1 - 12	12
1	Sum 1 Ch 1	Number 0 - 9	13 – 28	16
2	Sum 1 Ch 2	Numbers 10 - 19	29 – 54	26
3	Sum 1 Ch 3	Three Numbers	55 – 80	26
		Final Exam 1 A, 1 B, p73, p74		80
4	Sum 2 Ch 1	Numbers 20 – 29	81-104	24
5	Sum 2 Ch 2	Numbers 50 – 59	105-122	18
6	Sum 2 Ch 3	Numbers 30 – 39	123-144	22
		Final Exam 2A, 2 B, p 139, 140		64
7	Sum 3 Ch 1	Numbers 60 – 69	145-174	30
8	Sum 3 Ch 2	Numbers 90 – 99	175-190	16
9	Sum 3 Ch 3	Numbers 70 - 79	191 -212	22
		Final Exam 3 A, 3 B, p 207, 208		68
10	Sum 4 Ch 1	Three–Digit Numbers (+,-)	213 -232	20
11	Sum 4 Ch 2	Multiplication Numbers 1	233 -256	24
12	Sum 4 Ch 3	Division Numbers 1	257- 292	36
		Final Exam 4 A, 4 B, p 283, 284		80
13	Sum 5 Ch 1	Four-Digit Numbers (+,-)	293–314	22
14	Sum 5 Ch 2	Multiplication Numbers 2	315- 332	18
15	Sum 5 Ch 3	Division Numbers 2	333–348	16
16	Sum 5 Ch 4	4-Operations Numbers 2	349–376	28
		Final Exam 5 A, 5 B, p 370, 371		84
17	Sum 5 Ch 5	Many-Digit Numbers (+,-)	377 -401	25
18	Sum 5 Ch 6	Multiplication Numbers 3	402 -414	13
19	Sum 5 Ch 7	Division Numbers 3	415 -433	19
20	Sum 5 Ch 8	4-Operations with Numbers 3	434 -444	11
	20×3=60	Final Exam 6 A, 6 B, p 441 - 442		

Chapter 5, 6, 7, 8 has original solutions. Try to understand the solutions.
Rewrite problem in your notebook and try to do yourself.
After that check your answer with original solution.

Math Help for College

Table 1 Example

#	Summary = Sum	Grades	Average
	Diagnostic Test (DT)	68% = D+	
1	Sum 1 Final Exam 1 B (FE1)	74%= C	Av1 = (DT+ FE1) = (68 + 74)/2 = 71% = C
2	Sum 2 Final Exam 2 B (FE2)	80%=B	Av2=(FE1+FE2)/2 = (74+80)/2 = 77% C+
3	Sum 3 Final Exam 3 B (FE3)	86%=B+	AV3 = (FE1+FE2+FE3)/3=(74+80+86)/3=80 %=B
4	Sum 4 Final Exam 4 B (FE4)	92%=A	Av4=(FE1+FE2+FE3+FE4)/4= (74+80+86+92)/4 = = 83% B
5	Sum 5 Final Exam 5 B (FE5)	98%=A	Av5=((FE1+FE2+FE3+FE4+FE5)/5= = (74+80+86+92+98)/5=86B+
6	Sum5* Final Exam 6 B (FE6)		

Math Help for College

www.mathbooksbyirina.com

Students Attentions!!!!

Final Exam 1 A, Test 1 A, Review 1 A are different.

#	Summary = Sum	Name	Test A	Test B	Answers
	Sum 0	Diagnostic Test			
1	Sum 1	Final Exam 1 B (FE1B)	73	74	80
2	Sum 2	Final Exam 2 B (FE2B)	139	140	144
3	Sum 3	Final Exam 3 B (FE3B)	207	208	212
4	Sum 4	Final Exam 4 B (FE4B)	283	284	292
5	Sum 5	Final Exam 5 B (FE5B)	370	371	376

Table 3 Your Grades

	Summary = Sum	Grades	Average
	Diagnostic Test		
1)	Sum 1 Final Exam 1 B (FE1B)		Av1
2)	Sum 2 Final Exam 2 B (FE2B)		Av2
3)	Sum 3 Final Exam 3 B (FE3B)		Av3
4)	Sum 4 Final Exam 4 B (FE4B)		Av4
5)	Sum 5 Final Exam 5 B (FE5B)		Av5

Parents, Tutors, Students Middle Schools, High Schools, Colleges, University can help primary school students. Please read slowly math books to primary school students.

Students!

Students, you are smarter than a calculator!
Students, you are smarter than a computer!
Students, you need to believe in yourself.
Students, be patient and do your best.

Do not be afraid to make mistakes.
Tell yourself positive words.
I am young! I am smart!
I am strong! I can do anything!

Smart students use their brain!
Your brain is always with you!
A calculator is not always with you.
Mathematics is work for the brain.

Be open to new ideas in math!
Work hard and try your best!
Check answers and ask questions!
Practice! Practice! Practice!

Basic and Advanced Problems

You have two kinds of problems. There are basic problems: numbers 0 – 9, numbers 10 – 19, three numbers, numbers 20 – 29, numbers 50- 59, numbers 30-39, numbers 60 – 69, numbers 90-99, numbers 70-79. Also, you have advanced problems: three-digit numbers, four-digit numbers, many-digit numbers. Advanced problems help students improve attention, memory and logical thinking and get good knowledge that helps students in any job.

This book help people to learn math. If you never understood math and have not any knowledge in math you got chance to know this subject. If you were a small boy or girl and in school some problems you got very early for you; some problems were difficult for you. I have not enough time like teacher in school to explain all information. You need to want to know math and make your life better. You need to want to work with book and solve problems.

If you do not understand today, read again tomorrow and some day you feel better. I have some books in philosophy and I do not always understand material, but later I will be read again and again for understanding. Not every day we are smart in any reason. Some day you feel good, some day not. Some people got master degree in math for 5 years but I got master degree about 6 years. Nothing wrong, I know my math very good. You need to read book then makes notes, rewrite in your notebook information what you do not know or cannot remember.

Our brain has the same measurement in 18, in 19, in 20, and now in 21 centuries. We have much more information now than before; this is way we need work very slow; I am very slow person. I repeat rules in my book for you many times and if you work slow you will be known material. I played on piano for 5 years, from 4 grades to 8 grades, if I repeat many times my assignment, I played every day better and better. Do not rush, work with your speed. This is minimum material for your future. This book for adults, who never understand math in schools. This book is summary grades from 1 to 6, more important topics, rules, examples. If you want to be in college work with this book, read book, rules, work with examples, solve the problems, reviews, tests, final exams.

Psychology helps students.

Many schools want to use the calculator, but many colleges do not. This book helps you to be ready for college and teaches you how to work without a calculator, just use your brain. Do your best but try hard to solve problems. After that, check the answer.

To improve the student's attention, you need to rewrite problems from the book to their notebook, because 30% of our brain has a connection to the hands and showing all steps of solution gives you full credit. Students, be very, very, very, very patient with yourselves. You need to love yourself, care about your health, behavior, and education. Respect yourself, and respect others.

It would help to know some information from psychology.
Every 23 days, our physical condition changes.
Every 28 days, our emotional condition changes.
Every 33 days, our intellectual conditional changes.
We are smart, but not every day. You can make some mistakes.
Students understand material better if they are in good physical, emotional, and intellectual condition. When the students are ready to learn, the book can give students knowledge.

Students, we remember on average …
20 % of what we read, 30% of what we hear
40% of what we see with attention, 50% of what we say, read loudly
60% of what do, what we write, 70% of what we repeat rules
80% of what we practice, 90% of what we try to memorize
100% of what we read, hear, see, say, write, and teach others.

Students read the textbook, notes, tables, listen to the teacher, and watch with all your attention. Read loudly, write notes, and repeat many times to improve your memory. Practice, memorize and make it your goal to improve your logical thinking. You will soon be proud of yourself and happy with the result. Many students did it and you can do it too.

Just 10 problems from each lesson!

This book is the first part of mathematics where students learn how to do operations with numbers. This book contains lessons, rules, examples, exercises for practice, tests, and answers to exercises. This a book designed to help students improve their attention, memory, and logical thinking.

Students, you do not need to solve all of the problems, first check your knowledge. Students need to solve a minimum of 10 problems from each lesson to get good knowledge in math, a maximum of 20 problems. Grading scale: grade A = 10 problems solve correctly, grade B = 9 problems, grade C=8 problems, grade D = 7 problems, grade F = 0-6 problems.

If students taking the test got a grade an A, they pass the chapter and can move to next chapter or to the final exam. If students taking the test got a grade less than an A, they need to repeat rules and examples from each lesson. Then check their mistakes in the Test A and in the Final Exam A. After that, try to solve Test B or Final Exam B.

If students took the test and got a grade less than an A, never be upset, and just tell yourself positive things like "I tried my best!", "I just need to practice more" or "I need to ask more questions." Look at all of your grades on the test and find the average. For example, a student has 4 tests and Final Exam with the grades F=6, D=7, C=8, B=9, A = 10. The average is (6 + 7 + 8 + 9 + 10) and divided by 5, average would be 8 = B.

Math classes in elementary, middle, high school and college

1) Primary School: Basic four operations with whole numbers.
2) Middle School: College Mathematics, Prealgebra
3) 9th grade Beginning Algebra (1), Intermediate Algebra (2)
4) 10th grade Geometry and Trigonometry
5) 11th grade College Algebra (3) and Precalculus
6) 12th grade Calculus 1 and Calculus 2

BEAUTY OF MATHEMATICS!
IF THE LETTERS IS REPRESENTED AS THE NUMBERS!

A = 1	B = 2	C = 3
D = 4	E = 5	F = 6
G = 7	H = 8	I = 9
J = 10	K = 11	L = 12
M =13	N = 14	O = 15
P = 16	Q = 17	R = 18
S = 19	T = 20	U = 21
V = 22	W = 23	X =24
Y = 25	Z = 26	

WHAT EQUALS 100% IN LIFE?

KNOWLEDGE=11+14+15+23+12+5+4+7+5=96%
HARDWORK=8+1+18+4+23+15+18+11=98%
ATTITUDE=1+20+20+9+20+21+4+5=100%

IT'S THE ATTITUDE, HARDWORK, KNOWLEDGE, THAT WILL PUT YOU ON THE TOP.

Parents and Kids!

Parents, you need to help your kids. If kids feel your love and care they will be kind and smart. Encourage and hug kids. Tell them; I love you. I understand you. I forgive you. I proud of you. Forgive me, please. You can do anything. Try again. You did a great job. Thank you, dear. Be patient with your kids. Parents like God for kids until the age of 12. Kids of ages 13–19 need more help, because in this time they not adults. Parents, please open the internet and ask the question: how I can help a girl or boy of ages 13-19? It is very important when parents love and care about children. Children respect parents and be grateful to the them.

An average, adults work 8 hours a day. For kids, it is enough to work in school 6 hours a day from 8:00 am to 2:00 pm. After that kids have to play. Students have to learn different sciences and have different activities. Knowledge and health is very important for kids. Many doctors agree.

Schools and Kids!

Example for elementary School

1-4 grades, 8:00 am–1:30 pm, 30 mins break-play-time after each class

1st lesson 8:00-8:30 English	4th lesson 11:00–11:30 Science
2nd lesson 9:00-9:30 Spanish	5th lesson 12:00–12:30 Dance
3rd lesson 10:00–10:30 Math	6th lesson 1:00–1:30 Volleyball

Example for Middle School

5–8 grades, 8:00 am–1:45 pm,	15 mins break after each class
1st lesson 8:00–8:45 English	4th lesson 11:00–11:45 Geography
2nd lesson 9:00–9:45 Spanish	5th lesson 12:00–12:45 History
3rd lesson 10:00–10:45 Math	6th lesson 1:00-1:45 Dance

Example for High School

9–12 grades, 8:00 am–1:50 pm,	10 mins break after each class
1st lesson 8:00–8:50 English	4th lesson 11:00–11:50 Physics
2nd lesson 9:00–9:50 Spanish	5th lesson 12:00–12:50 Chemistry
3rd lesson 10:00 – 10:50 Math	6th lesson 1:00 – 1:50 Biology

Math Help

for College

Summary 1

Summary 1 Diagnosing Test
(25 minutes)

Name

Read directions carefully. Answers must be in the indicated place.

Show all steps for full credits. **Each problem is worth 5 points.**

Do only 1 operation per 1 step. **Solve 1 problem 5 minutes**

1) **Multiply** $\begin{array}{r} 807 \\ \underline{\times 504} \end{array}$

1)_______

2) **Divide** $245427 \div 403$

2)______

3) **Evaluate** $420420 \div 70 \times 2$

-

3)______

4) **Evaluate** $640640 - 320320 \div 4 \times 5$

4)______

5) **Evaluate** $180 \div 6 \times 2 - 66 \div 6 \times 3$

5)_______

How do you feel about Math?

a) Love b) Like c) Nothing d) Do not Like e) Difficult

Summary 1

Chapter 1

One – Digit

Numbers

Summary 1 Chapter 1 Contents
One-Digit Numbers
Lesson 1 – 8
Basic Problems

1)Work with lesson 1, read rules and examples, 2) Solve the problems from contents, 3) Check answers in the end of chapter 1, 4) Work with the next lessons 2, 3, 4, 5, ..., 5) Solve Test 1 A, 6) Grading Scale: A =10 points, B=9points, C=8 points, D=7 points, F=0-6 points, 7) If you solve 10 problems correctly, you get an A, 8) If have an A, you can move on to next chapter, 9) If your grade is lower than an A, work with the mistakes, then solve Test 1B, 10) When you finish all of the chapters, solve Final Exam in last chapter.

Lesson 1 Numbers from 0 to 9. Pretest # 1 - # 5
Solve the problems # 5 (a, b, c, d, e) and # 10 (a, b, c, d, e)
Lesson 2 Addition Numbers in row Review 1 A
Solve the problems # 5 (a, b, c, d, e) and #10 (a, b, c, d, e)
Property of Addition Review 1 B
Solve the problems # 5 (a, b, c, d, e) and #10 (a, b, c, d, e)

Lesson 3 Subtraction Numbers in row
Solve the problems # 5 (a, b, c, d, e) and #10 (a, b, c, d, e)
Lesson 4 Addition Numbers in column 1
Solve the problems # 5 (a, b, c, d, e) and # 10 (a, b, c, d, e)

Lesson 5 Addition Numbers in column 2
Solve the problems # 5 (a, b, c, d, e) and # 10 (a, b, c, d, e)
Lesson 6 Subtraction Numbers in column
Solve the problems # 5 (a, b, c, d, e) and # 10 (a, b, c, d, e)

<u>Lesson 7 Test for 50 minutes (10 problems)</u>
Solve Test 1 A # 1, 3, 5, 7, 9, 11, 13, 15, 17, 19
Solve Test 1 B # 1, 3, 5, 7, 9, 11, 13, 15, 17, 19
Lesson 8 Money Skills
Answers Chapter 1

Summary 1 Chapter 1 Lesson 1
One-Digit Numbers
Basic Problems
Numbers from 0 to 9

Count forward from 1 to 9 1, 2, 3, 4, 5, 6, 7, 8, 9
One, two, three, four, five, six, seven, eight, nine
First, second, third, fourth, fifth, sixth, seventh, eighth, ninth
Count backwards from 9 to 1 9, 8, 7, 6, 5, 4, 3, 2, 1
The smallest number is 0 The biggest number is 9
Equal means 5 = 5 Number zero is less than one 0 < 1
Less than (<) means

0 < 1	1 < 2	2 < 3	3 < 4	4 < 5
5 < 6	6 < 7	7 < 8	8 < 9	9 = 9

More than (>) means

9 > 8	8 > 7	7 > 6	6 > 5	5 > 4
4 > 3	3 > 2	2 > 1	1 > 0	0 = 0

To compare numbers, find what number greater or smaller or numbers equal 0 ? 3 0 < 3 3 ? 0 3 > 0

Put symbols <, > or = instead question mark

	a	b	c	d	e
# 1	0 ? 5	1 ? 5	2 ? 5	3 ? 5	4 ? 5
# 2	5 ? 5	6 ? 5	7 ? 5	8 ? 5	9 ? 5
# 3	0 ? 6	1 ? 6	2 ? 6	3 ? 6	4 ? 6
# 4	5 ? 6	6 ? 6	7 ? 6	8 ? 6	9 ? 6
# 5	0 ? 7	1 ? 7	2 ? 7	3 ? 7	4 ? 7
# 6	5 ? 7	6 ? 7	7 ? 7	8 ? 7	9 ? 7
# 7	0 ? 8	1 ? 8	2 ? 8	3 ? 8	4 ? 8
# 8	5 ? 8	6 ? 8	7 ? 8	8 ? 8	9 ? 8
# 9	0 ? 9	1 ? 9	2 ? 9	3 ? 9	4 ? 9
# 10	5 ? 9	6 ? 9	7 ? 9	8 ? 9	9 ? 9

Summary 1 Chapter 1 Lesson 2

Addition numbers in row

The first operation with numbers is addition +
To add numbers put numbers together: 2 + 3 = 5
How many in all or how many together? If put together 2 and 3 results is 5. If you need to add 5 + 3 you have to count after 5 forward 3 times That should be 6, 7, 8, that is why 5 + 3 = 8. The sum of a number and zero is the same as the number: 5 + 0 = 5 0 + 5 = 5

Add numbers. Review 1 A

	a	b	c	d	e
# 1	0 + 0	1 + 0	2 + 0	3 + 0	4 + 0
# 2	5 + 0	6 + 0	7 + 0	8 + 0	9 + 0
# 3	0 + 1	1 + 1	2 + 1	3 + 1	4 + 1
# 4	5 + 1	6 + 1	7 + 1	8 + 1	1 + 8
# 5	0 + 2	1 + 2	2 + 2	3 + 2	4 + 2
# 6	5 + 2	6 + 2	7 + 2	2 + 7	1 + 7
# 7	0 + 3	1 + 3	2 + 3	3 + 3	4 + 3
# 8	5 + 3	6 + 3	0 + 5	1 + 5	2 + 5
# 9	0 + 4	1 + 4	2 + 4	3 + 4	4 + 4
# 10	5 + 4	3 + 5	4 + 5	0 + 6	2 + 6

Property of addition. Review 1 B

Two numbers can add in any order. The sum wills the same. 2+3=5 3+2=5. If you need to add 3 + 6, easy to start from bigger number, you need to start after 6 by count forward 3 times, that should be 7, 8, 9. That why 6 + 3 = 9
3 + 6 = 6 + 3 = 9

Add numbers

	a	b	c	d	e
# 1	0 + 1	1 + 1	2 + 1	3 + 1	4 + 1
# 2	5 + 1	6 + 1	7 + 1	8 + 1	1 + 8
# 3	0 + 2	1 + 2	2 + 2	3 + 2	4 + 2
# 4	5 + 2	6 + 2	7 + 2	2 + 7	1 + 7
# 5	0 + 3	1 + 3	2 + 3	3 + 3	4 + 3
# 6	5 + 3	6 + 3	6 + 1	6 + 2	6 + 3
# 7	0 + 4	1 + 4	2 + 4	3 + 4	4 + 4
# 8	0 + 5	1 + 5	2 + 5	3 + 5	4 + 5
# 9	0 + 6	1 + 6	2 + 6	3 + 6	0 + 8
# 10	0 + 7	1 + 7	2 + 7	1 + 8	5 + 4

Summary 1 Chapter 1 Lesson 3
Subtraction numbers in row

Second operation with numbers is subtraction –
To subtract numbers, take away something: 5 – 2 = 3

How many are left? 3 are left

If you need to subtract 5 – 3 you have to count backward from 5 three times 4, 3, 2
This is why 5 – 3 = 2
Any number subtract zero result is the same number: 5 – 0 = 5
Any number subtract the same number result is zero: 5 – 5 = 0

Subtraction numbers in row
9 - 2 = 7
Count from 9 backward 2 times 8, 7
Subtract numbers

	a	b	c	d	e
# 1	0 – 0	1 – 0	2 – 0	3 - 0	4 – 0
# 2	5 – 0	6 – 0	7 – 0	8 – 0	9 - 0
# 3	1 – 1	2 – 1	3 – 1	4 – 1	5 - 1
# 4	6 – 1	7 – 1	8 – 1	9 – 1	8 – 7
# 5	2 – 2	3 – 2	4 – 2	5 – 2	6 – 2
# 6	7 – 2	8 – 2	9 – 2	9 – 8	9 - 7
# 7	3 – 3	4 – 3	5 – 3	6 – 3	7 – 3
# 8	4 – 4	5 – 4	6 – 4	7 – 4	8 – 4
# 9	5 – 5	6 – 5	7 – 5	8 – 5	9 – 5
# 10	6 – 6	7 – 6	8 – 6	9 – 6	9 - 9

Summary 1 Chapter 1 Lesson 4
Addition numbers in column 1

7	5
+2	+4
9	9

Add numbers

	a	b	c	d	e
# 1	0	1	2	3	4
	+0	+ 0	+0	+0	+0
# 2	5	6	7	8	9
	+ 0	+ 0	+ 0	+ 0	+0
# 3	0	1	2	3	4
	+ 1	+ 1	+ 1	+ 1	+1
# 4	5	6	7	8	1
	+ 1	+ 1	+ 1	+ 1	+8
# 5	0	1	2	3	4
	+ 2	+ 2	+ 2	+ 2	+ 2
# 6	5	6	7	2	1
	+ 2	+ 2	+ 2	+ 7	+7
# 7	0	1	2	3	4
	+ 3	+ 3	+ 3	+ 3	+ 3
# 8	5	6	0	1	2
	+ 3	+ 3	+ 5	+ 5	+ 5
# 9	0	1	2	3	4
	+ 4	+ 4	+ 4	+ 4	+ 4
# 10	5	3	4	0	2
	+ 4	+ 5	+ 5	+ 6	+ 6

Summary 1 Chapter 1 Lesson 5
Addition numbers in column 2

8	6
+ 1	+3
9	9

Add numbers

	a	b	c	d	e
# 1	1 + 0	1 + 1	1 + 2	1 + 3	1 + 4
# 2	1 + 5	1 + 6	1 + 7	1 + 8	8 + 1
# 3	2 + 0	2 + 1	2 + 2	2 + 3	2 + 4
# 4	2 + 5	2 + 6	2 + 7	7 + 2	7 + 1
# 5	3 + 0	3 + 1	3 + 2	3 + 3	3 + 4
# 6	3 + 5	3 + 6	1 + 6	2 + 6	3 + 6
# 7	4 + 0	4 + 1	4 + 2	4 + 3	4 + 4
# 8	5 + 0	5 + 1	5 + 2	5 + 3	5 + 4
# 9	6 + 0	6 + 1	6 + 2	6 + 3	8 + 0
# 10	7 + 0	7 + 1	7 + 2	8 + 1	4 + 5

Summary 1 Chapter 1 Lesson 6
Subtraction numbers in column

6	7
-2	- 2
4	5

Subtract numbers

	a	b	c	d	e
# 1	0	1	2	3	4
	- 0	- 0	- 0	- 0	- 0
# 2	5	6	7	8	9
	- 0	- 0	- 0	- 0	- 0
# 3	1	2	3	4	5
	- 1	- 1	- 1	- 1	- 1
# 4	6	7	8	9	8
	- 1	- 1	- 1	- 1	- 7
# 5	2	3	4	5	6
	- 2	- 2	- 2	- 2	- 2
# 6	7	8	9	9	9
	- 2	- 2	- 2	- 8	- 7
# 7	3	4	5	6	7
	- 3	- 3	- 3	- 3	- 3
# 8	4	5	6	7	8
	- 4	- 4	- 4	- 4	- 4
# 9	5	6	7	8	9
	- 5	- 5	- 5	- 5	- 5
# 10	6	7	8	9	9
	- 6	- 6	- 6	- 6	- 9

Summary 1 Chapter 1 Lesson 7
Test 1 A
Add or Subtract numbers in row

# 1	1 + 8	# 6	4 – 4
# 2	2 + 2	# 7	6 – 3
# 3	7 + 2	# 8	9 – 2
# 4	2 + 4	# 9	5 – 2
# 5	6 + 2	# 10	9 – 1

Add or Subtract numbers in column

# 11	3 + 6	# 12	2 + 4	# 13	4 + 3	# 14	4 + 4	# 15	4 + 5
# 16	9 -3	# 17	5 - 3	# 18	8 - 2	# 19	4 - 2	# 20	8 - 1

Summary 1 Chapter 1 Lesson 7
Test 1 B
Add or Subtract numbers in row

1 5 + 4

2 0 + 6

3 6 + 1

4 2 + 6

5 6 + 3

6 8 – 3

7 4 – 3

8 7 – 2

9 3 – 2

10 7 – 0

Add or Subtract numbers in column

11 $\begin{array}{r} 7 \\ \underline{+1} \end{array}$ # 12 $\begin{array}{r} 9 \\ \underline{+0} \end{array}$ # 13 $\begin{array}{r} 2 \\ \underline{+7} \end{array}$ # 14 $\begin{array}{r} 8 \\ \underline{+1} \end{array}$ # 15 $\begin{array}{r} 0 \\ \underline{+9} \end{array}$

16 $\begin{array}{r} 7 \\ \underline{-3} \end{array}$ # 17 $\begin{array}{r} 3 \\ \underline{-3} \end{array}$ # 18 $\begin{array}{r} 6 \\ \underline{-2} \end{array}$ # 19 $\begin{array}{r} 2 \\ \underline{-2} \end{array}$ # 20 $\begin{array}{r} 6 \\ \underline{-0} \end{array}$

Summary 1 Chapter 1 Lesson 8
Money Skills

1 penny = 1 cent

1 nickel = 5 cents = 1c + 1c + 1c + 1c + 1c

1 dime = 10 cents = 5c + 5c

How many ones or pennies?
2 pennies = 1c + 1c
3 pennies = 1c + 1c + 1c
4 pennies = 1c + 1c + 1c + 1c
6 pennies = 5c + 1c
7 pennies = 5c + 1c + 1c
8 pennies = 5c + 1c + 1c + 1c
9 pennies = 5c + 1c + 1c + 1c + 1c

How many tens or dimes?
1 ten = 10 ones
2 tens = 20 ones = 10c + 10c
3 tens = 30 ones = 10c + 10c + 10c
4 tens = 40 ones = 10c + 10c + 10c + 10c
5 tens = 50 ones = 10c + 10c + 10c + 10c + 10c
6 tens = 60 ones = 10c + 10c + 10c + 10c + 10c + 10c

Skip Counting. Write the missing numbers.
Count by 2 (+2)   1, 3, 5, ?, ?.

Summary 1 Answers Chapter 1

Pretest 1 1) 406,728 2) 609 3) 12012 4) 240,240 5) 27

Lesson 1

	a	b	c	d	e
# 1	0 < 5	1 < 5	2 < 5	3 < 5	4 < 5
# 2	5 = 5	6 > 5	7 > 5	8 > 5	9 > 5
# 3	0 < 6	1 < 6	2 < 6	3 < 6	4 < 6
# 4	5 < 6	6 = 6	7 > 6	8 > 6	9 > 6
# 5	0 < 7	1 < 7	2 < 7	3 < 7	4 < 7
# 6	5 < 7	6 < 7	7 = 7	8 > 7	9 > 7
# 7	0 < 8	1 < 8	2 < 8	3 < 8	4 < 8
# 8	5 < 8	6 < 8	7 < 8	8 = 8	9 > 8
# 9	0 < 9	1 < 9	2 < 9	3 < 9	4 < 9
# 10	5 < 9	6 < 9	7 < 9	8 < 9	9 = 9

Lesson 2 Review 1 A

	a	b	c	d	e
# 1	0	1	2	3	4
# 2	5	6	7	8	9
# 3	1	2	3	4	5
# 4	6	7	8	9	9
# 5	2	3	4	5	6
# 6	7	8	9	9	8
# 7	3	4	5	6	7
# 8	8	9	5	6	7
# 9	4	5	6	7	8
# 10	9	8	9	6	8

Lesson 2 Review 1 B

	a	b	c	d	e
# 1	1	2	3	4	5
# 2	6	7	8	9	9
# 3	2	3	4	5	6
# 4	7	8	9	9	8
# 5	3	4	5	6	7
# 6	8	9	7	8	9
# 7	4	5	6	7	8
# 8	5	6	7	8	9
# 9	6	7	8	9	8
# 10	7	8	9	9	9

Summary 1 Answers Chapter 1

Lesson 3

	a	b	c	d	e
# 1	0	1	2	3	4
# 2	5	6	7	8	9
# 3	0	1	2	3	4
# 4	5	6	7	8	1
# 5	0	1	2	3	4
# 6	5	6	7	1	2
# 7	0	1	2	3	4
# 8	0	1	2	3	4
# 9	0	1	2	3	4
# 10	0	1	2	3	0

Lesson 4

	a	b	c	d	e
# 1	0	1	2	3	4
# 2	5	6	7	8	9
# 3	1	2	3	4	5
# 4	6	7	8	9	9
# 5	2	3	4	5	6
# 6	7	8	9	9	8
# 7	3	4	5	6	7
# 8	8	9	5	6	7
# 9	4	5	6	7	8
# 10	9	8	9	6	8

Lesson 5

	a	b	c	d	e
# 1	1	2	3	4	5
# 2	6	7	8	9	9
# 3	2	3	4	5	6
# 4	7	8	9	9	8
# 5	3	4	5	6	7
# 6	8	9	7	8	9
# 7	4	5	6	7	8
# 8	5	6	7	8	9
# 9	6	7	8	9	8
# 10	7	8	9	9	9

Summary 1 Answers Chapter 1

Lesson 6

	a	b	c	d	e
# 1	0	1	2	3	4
# 2	5	6	7	8	9
# 3	0	1	2	3	4
# 4	5	6	7	8	1
# 5	0	1	2	3	4
# 6	5	6	7	1	2
# 7	0	1	2	3	4
# 8	0	1	2	3	4
# 9	0	1	2	3	4
# 10	0	1	2	3	0

Lesson 7
Test 1 A

# 1	9	# 2	4	# 3	9	# 4	6	# 5	8
# 6	0	# 7	3	# 8	7	# 9	3	# 10	8
# 11	9	# 12	6	# 13	7	# 14	8	# 15	9
# 16	6	# 17	2	# 18	6	# 19	2	# 20	7

Lesson 7
Test 1 B

# 1	9	# 2	6	# 3	7	# 4	8	# 5	9
# 6	5	# 7	1	# 8	5	# 9	1	# 10	7
# 11	8	# 12	9	# 13	9	# 14	9	# 15	9
# 16	4	# 17	0	# 18	4	# 19	0	# 20	6

<u>Summary 1</u>

Chapter 2

Two – Digit

Numbers

Summary 1 Chapter 2 Contents
Two-Digit Numbers
Lesson 1 - 16
Basic Problems

Students you need to buy counting sticks for calculations (100)

Work with lesson 1, read rules and examples, 2) Solve the problems from contents, 3) Check answers in the end of chapter 1, 4) Work with the next lessons 2, 3, 4, 5, …, 5) Solve Test 1 A, 6) Grading Scale: A =10 points, B=9points, C=8 points, D=7 points, F=0-6 points, 7) If you solve 10 problems correctly, you get an A, 8) If have an A, you can move on to next chapter, 9) If your grade is lower than an A, work with the mistakes, then solve Test 1B, 10) When you finish all of the chapters, solve Final Exam in last chapter. Lesson = L

Summary 1 Chapter 2 Lesson 1
One-digit numbers
One-digit numbers have one digit
This is numbers from 0 to 9
0, 1, 2, 3, 4, 5, 6, 7, 8, 9

Add or Subtract numbers in column

# 1	3 +6	# 2	2 + 4	# 3	4 + 3	# 4	4 + 4	# 5	4 +5
# 6	9 -3	# 7	5 - 3	# 8	8 - 2	# 9	4 -2	# 10	8 - 1

Count money: 50 cents, 1 dollar
1 cent = 1 penny
1 nickel = 5 cent
1 dime = 10 cent
1 quarter = 25 cent

Add or Subtract numbers in column

# 11	\$7 + \$1	# 12	\$9 + \$0	# 13	\$2 + \$7	# 14	\$8 + \$1	# 15	\$ 0 +\$ 9
# 16	\$7 -\$3	# 17	\$3 - \$3	# 18	\$6 - \$2	# 19	\$2 - \$2	# 20	\$6 - \$0

Summary 1 Chapter 2 Lesson 2
Two-digit numbers

Two - digit numbers have two digits
This is numbers from 10 to 99
Digit on the left are called tens
Digit on the right are called ones
Digits 0, 1, 2, 3, 4, 5, 6, 7, 8, 9

Ten, eleven, twelve, thirteen, fourteen,
fifteen, sixteen, seventeen, eighteen, nineteen

Read numbers 10 to 19: 10, 11, 12, 13, 14, 15, 16, 17, 18, 19
Read numbers 20 through 29: 20, 21, 22, 23, 24, 25, 26, 27, 28, 29
Read numbers 30 through 39: 30, 31, 32, 33, 34, 35, 36, 37, 38, 39
Read numbers 40 through 49: 40, 41, 42, 43, 44, 45, 46, 47, 48, 49
Read numbers 50 through 59: 50, 51, 52, 53, 54, 55, 56, 57, 58, 59
Read numbers 60 through 69: 60, 61, 62, 63, 64, 65, 66, 67, 68, 69
Read numbers 70 through 79: 70, 71, 72, 73, 74, 75, 76, 77, 78, 79
Read numbers 80 through 89: 80, 81, 82, 83, 84, 85, 86, 87, 88, 89
Read numbers 90 through 99: 90, 91, 92, 93, 94, 95, 96, 97, 98, 99
Rewrite and Memorize numbers from 10 to 99

Three-digit numbers

Three-digit numbers content three digits. This is numbers 100, ... 999
Least three-digit number: 100. Great three-digit number: 999
Digits for three-digit numbers from right to the left: Ones, tens, and hundreds: 238
Counting numbers are numbers that use for calculations 1, 2, 3, . . . (+1)
Odd numbers are 1, 3, 5, 7, 9, 11, 13, 15, 17, 19... (+ 2)
Even numbers are 0, 2, 4, 6, 8, 10, 12, 14, 16, 18, 20 ... (+ 2)

Remember

$10 = 9 + 1 = 8 + 2 = 7 + 3 = 6 + 4 = 5 + 5$
$11 = 9 + 2 = 8 + 3 = 7 + 4 = 6 + 5$
$12 = 9 + 3 = 8 + 4 = 7 + 5 = 6 + 6$
$13 = 9 + 4 = 8 + 5 = 7 + 6$
$14 = 9 + 5 = 8 + 6 = 7 + 7$
$15 = 9 + 6 = 8 + 7$
$16 = 9 + 7 = 8 + 8$
$17 = 9 + 8$
$18 = 9 + 9$
$19 = 10 + 9$

Summary 1 Chapter 2 Lesson 3
Comparison numbers

To compare numbers, find what number greater or smaller or numbers are equal. One-digit numbers always less than two-digit numbers 9 < 11 Two-digit numbers always greater than one-digit numbers 19 > 8

Put symbols < or > or = instead of question mark

	a	b	c	d	e
# 1	0 ? 10	10 ? 1	2 ? 10	10 ? 3	4 ? 10
# 2	0 ? 11	11 ? 1	2 ? 11	11 ? 3	11 ? 4
# 3	0 ? 12	12 ? 1	2 ? 12	3 ? 12	12 ? 4
# 4	13 ? 0	1 ? 13	13 ? 2	13 ? 3	4 ? 13
# 5	14 ? 0	14 ? 1	2 ? 14	3 ? 14	4 ? 14
# 6	15 ? 0	1 ? 15	15 ? 2	3 ? 15	15 ? 4
# 7	0 ? 16	1 ? 16	16 ? 2	16 ? 3	4 ? 16
# 8	17 ? 0	1 ? 17	17 ? 2	17 ? 3	4 ? 17
# 9	18 ? 0	1 ? 18	2 ? 18	18 ? 3	4 ? 18
# 10	19 ? 0	1 ? 19	19 ? 2	19 ? 3	4 ? 19
# 11	5 ? 10	10 ? 6	7 ? 10	10 ? 8	9 ? 10
# 12	11 ? 5	6 ? 11	11 ? 7	11? 8	9 ? 11
# 13	5 ? 12	6 ? 12	12 ? 7	8 ? 12	9 ? 12
# 14	13 ? 5	6 ? 13	13 ? 7	13 ? 8	13 ? 9
# 15	5 ? 14	14 ? 6	7 ? 14	8 ? 14	9 ? 14
# 16	15 ? 5	6 ? 15	15 ? 7	8 ? 15	15 ? 9
# 17	5 ? 16	16 ? 6	7 ? 16	8 ? 16	16 ? 9
# 18	17 ? 5	17 ? 6	7 ? 17	17 ? 8	17 ? 9
# 19	5 ? 18	18 ? 6	7 ? 18	18 ? 8	9 ? 18
# 20	19 ? 5	6 ? 19	19 ? 7	8 ? 19	19 ? 9

Summary 1 Chapter 2 Lesson 4
Comparison two-digit numbers

To compare numbers, find what number greater or smaller or numbers equal. Compare the tens and then compare the ones 12 < 18   18 > 12

Put symbols < or > or = instead of question mark

	a	b	c	d	e
# 1	10 ? 10	11 ? 11	12 ? 12	11 ? 10	12 ? 10
# 2	10 ? 11	12 ? 11	12 ? 13	11 ? 12	13 ? 12
# 3	12 ? 10	11 ? 13	14 ? 12	13 ? 11	12 ? 14
# 4	10 ? 13	14 ? 11	15 ? 12	11 ? 14	12 ? 15
# 5	14 ? 10	11 ? 15	16 ? 12	15 ? 11	12 ? 16
# 6	10 ? 15	16 ? 11	12 ? 17	11 ? 16	17 ? 12
# 7	10 ? 16	17 ? 11	12 ? 18	11 ? 17	18 ? 12
# 8	17 ? 10	11 ? 18	19 ? 12	18 ? 11	12 ? 19
# 9	10 ? 18	11 ? 19	19 ? 10	18 ? 10	10 ? 19
# 10	11 ? 17	18 ? 15	12 ? 16	15 ? 18	16 ? 12
# 11	13 ? 13	14 ? 14	15 ? 15	14 ? 13	13 ? 12
# 12	13 ? 14	15 ? 14	15 ? 16	14 ? 15	16 ? 15
# 13	15 ? 13	14 ? 16	17 ? 15	16 ? 14	15 ? 17
# 14	13 ? 16	17 ? 14	15 ? 18	14 ? 17	18 ? 15
# 15	17 ? 13	14 ? 18	15 ? 19	18 ? 14	19 ? 15
# 16	13 ? 18	19 ? 14	13 ? 19	14 ? 19	19 ? 13
# 17	16 ? 16	17 ? 17	18 ? 18	17 ? 16	18 ? 13
# 18	16 ? 17	18 ? 17	18 ? 19	17 ? 18	19 ? 18
# 19	18 ? 16	17 ? 19	19 ? 19	19 ? 17	16 ? 18
# 20	16 ? 19	19 ? 18	15 ? 17	18 ? 19	17 ? 15

Summary 1 Chapter 2 Lesson 5

Addition numbers in row 1
5 + 9 = 9 + 5 = 14

Count from 9 forward 5 times 10, 11, 12, 13, 14
4 + 10 = 10 + 4 = 14
Count from 10 forward 4 times 11, 12, 13, 14

Add numbers

	a	b	c	d	e
# 1	0 + 1	1 + 1	2 + 1	3 + 1	4 + 1
# 2	5 + 1	6 + 1	7 + 1	8 + 1	9 + 1
# 3	0 + 2	1 + 2	2 + 2	3 + 2	4 + 2
# 4	5 + 2	6 + 2	7 + 2	8 + 2	9 + 2
# 5	0 + 3	1 + 3	2 + 3	3 + 3	4 + 3
# 6	5 + 3	6 + 3	7 + 3	8 + 3	9 + 3
# 7	0 + 4	1 + 4	2 + 4	3 + 4	4 + 4
# 8	5 + 4	6 + 4	7 + 4	8 + 4	9 + 4
# 9	0 + 5	1 + 5	2 + 5	3 + 5	4 + 5
# 10	5 + 5	6 + 5	7 + 5	8 + 5	9 + 5
# 11	0 + 6	1 + 6	2 + 6	3 + 6	4 + 6
# 12	5 + 6	6 + 6	7 + 6	8 + 6	9 + 6
# 13	0 + 7	1 + 7	2 + 7	3 + 7	4 + 7
# 14	5 + 7	6 + 7	7 + 7	8 + 7	9 + 7
# 15	0 + 8	1 + 8	2 + 8	3 + 8	4 + 8
# 16	5 + 8	6 + 8	7 + 8	8 + 8	9 + 8
# 17	0 + 9	1 + 9	2 + 9	3 + 9	4 + 9
# 18	5 + 9	6 + 9	7 + 9	8 + 9	9 + 9
# 19	0 + 10	1 + 10	2 + 10	3 + 10	4 + 10
# 20	5 + 10	6 + 10	7 + 10	8 + 10	9 + 10

Summary 1 Chapter 2 Lesson 6

Addition numbers in row 2

7 + 12 = 12 + 7 = 10 + 2 + 7 = 10 + 9 = 19

6 + 12 = 12 + 6 = 10 + 2 + 6 = 10 + 8 = 18

Add numbers

	a	b	c	d	e
# 1	8 + 0	8 + 1	8 + 2	8 + 3	8 + 4
# 2	8 + 5	8 + 6	8 + 7	8 + 8	8 + 9
# 3	9 + 0	9 + 1	9 + 2	9 + 3	9 + 4
# 4	9+ 5	9 + 6	9 + 7	9 + 8	9 + 9
# 5	10 + 0	10 + 1	10 + 2	10 + 3	10 + 4
# 6	10 + 5	10 + 6	10 + 7	10 + 8	10 + 9
# 7	11 + 0	11 + 1	11 + 2	11 + 3	11 + 4
# 8	11 + 5	11 + 6	11 + 7	11 + 8	8 + 11
# 9	12 + 0	12 + 1	12 + 2	12 + 3	12 + 4
# 10	12 + 5	12 + 6	12 + 7	7 + 12	6 + 12
# 11	13 + 0	13 + 1	13 + 2	13 + 3	13 + 4
# 12	13 + 5	13 + 6	4 + 13	5 + 13	6 + 13
# 13	14 + 0	14 + 1	14 + 2	14 + 3	14 + 4
# 14	14 + 5	2 + 14	3 + 14	4 + 14	5 + 14
# 15	15 + 0	15 + 1	15 + 2	15 + 3	15 + 4
# 16	0 + 15	1 + 15	2 + 15	3 + 15	4 + 15
# 17	16 + 0	16 + 1	16 + 2	16 + 3	19 + 0
# 18	0 + 16	1 + 16	2 + 16	3 + 16	0 + 19
# 19	17 + 0	17 + 1	17 + 2	18 + 0	18 + 1
# 20	0 + 17	1 + 17	2 + 17	0 + 18	1 + 18

Summary 1 Chapter 2 Lesson 7

Subtraction numbers in row

14 – 4 = 10
Count from 14 backward 4 times 13, 12, 11, 10
18 – 4 = 14
Count from 18 backward 4 times 17, 16, 15, 14

Subtract numbers

	a	b	c	d	e
# 1	10 - 0	10 - 1	10 - 2	10 - 3	10 - 4
# 2	10 - 5	10 - 6	10 - 7	10 - 8	10 - 9
# 3	11 - 0	11 - 1	11 - 2	11 - 3	11 - 4
# 4	11 - 5	11 - 6	11 - 7	11 - 8	11 - 9
# 5	12 - 0	12 - 1	12 - 2	12 - 3	12 - 4
# 6	12 - 5	12 - 6	12 - 7	12 - 8	12 - 9
# 7	13 - 0	13 - 1	13 - 2	13 - 3	13 - 4
# 8	13 - 5	13 - 6	13 - 7	13 - 8	13 - 9
# 9	14 - 0	14 - 1	14 - 2	14 - 3	14 - 4
# 10	14 - 5	14 - 6	14 - 7	14 - 8	14 - 9
# 11	15 – 0	15 – 1	15 - 2	15 - 3	15 - 4
# 12	15 - 5	15 - 6	15 - 7	15 - 8	15 - 9
# 13	16 - 0	16 - 1	16 - 2	16 - 3	16 - 4
# 14	16 - 5	16 - 6	16 - 7	16 - 8	16 - 9
# 15	17 - 0	17 - 1	17 - 2	17 - 3	17 - 4
# 16	17 - 5	17 - 6	17 - 7	17 - 8	17 - 9
# 17	18 - 0	18 - 1	18 - 2	18 - 3	18 - 4
# 18	18 - 5	18 - 6	18 - 7	18 - 8	18 - 9
# 19	19 - 0	19 - 1	19 - 2	19 - 3	19 - 4
# 20	19 - 5	19 - 6	19 - 7	19 - 8	19 - 9

Summary 1 Chapter 2 Lesson 8

Addition and Subtraction in column
Number 10 and Number 11

6	11
+4	-2
10	9

Add or Subtract numbers

	a	b	c	d
# 1	10 + 0	10 - 1	5 + 5	10 - 6
# 2	9 + 1	10 - 2	4 + 6	10 - 7
# 3	8 + 2	10 - 3	3 + 7	10 - 8
# 4	7 + 3	10 - 4	2 + 8	10 - 9
# 5	6 + 4	10 - 5	1 + 9	10 - 10
# 6	10 + 1	11 - 1	5 + 6	11 - 6
# 7	9 + 2	11 - 2	4 + 7	11 - 7
# 8	8 + 3	11 - 3	3 + 8	11 - 8
# 9	7 + 4	11 - 4	2 + 9	11 - 9
# 10	6 + 5	11 - 5	1 + 10	11 - 10

Summary 1 Chapter 2 Lesson 9
Addition and Subtraction in column
Number 12 and Number 13

9	13
+3	-7
12	6

Add or Subtract numbers

	a	b	c	d
# 1	12 + 0	12 - 0	7 + 5	12 - 5
# 2	11 + 1	12 - 1	6 + 6	12 - 6
# 3	10 + 2	12 - 2	5 + 7	12 - 7
# 4	9 + 3	12 - 3	4 + 8	12 - 8
# 5	8 + 4	12 - 4	3 + 9	12 - 9
# 6	13 + 0	13 - 0	8 + 5	13 - 5
# 7	12 + 1	13 - 1	7 + 6	13 - 6
# 8	11 + 2	13 - 2	6 + 7	13 - 7
# 9	10 + 3	13 - 3	5 + 8	13 - 8
# 10	9 + 4	13 - 4	4 + 9	13 - 9

Summary 1 Chapter 2 Lesson 10

Addition and Subtraction in column

Number 14

9	14
+ 5	-9
14	5

Add or Subtract numbers

	a	b	c	d
# 1	14	14	9	14
	+ 0	- 0	+ 5	- 5
# 2	13	14	8	14
	+ 1	- 1	+ 6	- 6
# 3	12	14	7	14
	+ 2	- 2	+ 7	- 7
# 4	11	14	6	14
	+ 3	- 3	+ 8	- 8
# 5	10	14	5	14
	+ 4	- 4	+ 9	- 9
# 6	4	14	12	12
	+ 10	- 10	+ 0	- 0
# 7	3	14	12	12
	+ 11	- 11	+ 0	- 0
# 8	2	14	11	12
	+ 12	- 12	+ 1	- 9
# 9	1	14	10	12
	+ 13	- 13	+ 2	- 10
# 10	0	14	12	12
	+ 14	- 14	- 12	- 11

Summary 1 Chapter 2 Lesson 11
Addition and Subtraction in column
Number 15

$$\begin{array}{r} 9 \\ +6 \\ \hline 15 \end{array} \qquad \begin{array}{r} 15 \\ -9 \\ \hline 6 \end{array}$$

Add or Subtract numbers

	a	b	c	d
# 1	15 + 0	15 - 0	10 + 5	15 - 5
# 2	14 + 1	15 - 1	9 + 6	15 - 6
# 3	13 + 2	15 - 2	8 + 7	15 - 7
# 4	12 + 3	15 - 3	7 + 8	15 - 8
# 5	11 + 4	15 - 4	6 + 9	15 - 9
# 6	5 + 10	15 - 10	13 + 0	13 - 10
# 7	4 + 11	15 - 11	12 + 1	13 - 11
# 8	3 + 12	15 - 12	11 + 2	13 - 12
# 9	2 + 13	15 - 13	10 + 3	13 - 13
# 10	1 + 14	15 - 14	0 + 13	15 - 15

Summary 1 Chapter 2 Lesson 12
Addition and Subtraction in column
Number 16

9	16
+7	-9
16	7

Add or Subtract numbers

	a	b	c	d
# 1	16 + 0	16 - 0	11 + 5	16 - 5
# 2	15 + 1	16 - 1	10 + 6	16 - 6
# 3	14 + 2	16 - 2	9 + 7	16 - 7
# 4	13 + 3	16 - 3	8 + 8	16 - 8
# 5	12 + 4	16 - 4	7 + 9	16 - 9
# 6	6 + 10	16 - 10	1 + 15	16 - 5
# 7	5 + 11	16 - 11	0 + 16	16 - 6
# 8	4 + 12	16 - 12	9 + 7	16 - 7
# 9	3 + 13	16 - 13	8 + 8	16 - 8
# 10	2 + 14	16 - 14	7 + 9	16 - 9

Summary 1 Chapter 2 Lesson 13
Addition and Subtraction in column
Number 17

9	17
+8	-9
17	8

Add or Subtract numbers

	a	b	c	d
# 1	17 + 0	17 - 0	12 + 5	17 - 5
# 2	16 + 1	17 - 1	11 + 6	17 - 6
# 3	15 + 2	17 - 2	10 + 7	17 - 7
# 4	14 + 3	17 - 3	9 + 8	17 - 8
# 5	13 + 4	17 - 4	8 + 9	17 - 9
# 6	7 + 10	17 - 10	2 + 15	17 - 5
# 7	6 + 11	17 - 11	1 + 16	17 - 6
# 8	5 + 12	17 - 12	0 + 17	17 - 7
# 9	4 + 13	17 - 13	9 + 8	17 - 8
# 10	3 + 14	17 - 14	8 + 9	17 - 9

Summary 1 Chapter 2 Lesson 14
Addition and Subtraction in column
Number 18

9	18
+9	-9
18	9

Add or Subtract numbers

	a	b	c	d
# 1	18 + 0	18 - 0	13 + 5	18 - 5
# 2	17 + 1	18 - 1	12 + 6	18 - 6
# 3	16 + 2	18 - 2	11 + 7	18 - 7
# 4	15 + 3	18 - 3	10 + 8	18 - 8
# 5	14 + 4	18 - 4	9 + 9	18 - 9
# 6	8 + 10	18 - 10	3 + 15	18 - 5
# 7	7 + 11	18 - 11	2 + 16	18 - 6
# 8	6 + 12	18 - 12	1 + 17	18 - 7
# 9	5 + 13	18 - 13	0 + 18	18 - 8
# 10	4 + 14	18 - 14	9 + 9	18 - 9

Summary 1 Chapter 2 Lesson 15
Addition and Subtraction in column
Number 19

$$\begin{array}{r} 10 \\ +9 \\ \hline 19 \end{array} \qquad \begin{array}{r} 19 \\ -10 \\ \hline 9 \end{array}$$

Add or Subtract numbers

	a	b	c	d
# 1	$\begin{array}{r} 19 \\ +\ 0 \\ \hline \end{array}$	$\begin{array}{r} 19 \\ -\ 0 \\ \hline \end{array}$	$\begin{array}{r} 14 \\ +\ 5 \\ \hline \end{array}$	$\begin{array}{r} 19 \\ -\ 5 \\ \hline \end{array}$
# 2	$\begin{array}{r} 18 \\ +\ 1 \\ \hline \end{array}$	$\begin{array}{r} 19 \\ -\ 1 \\ \hline \end{array}$	$\begin{array}{r} 13 \\ +\ 6 \\ \hline \end{array}$	$\begin{array}{r} 19 \\ -\ 6 \\ \hline \end{array}$
# 3	$\begin{array}{r} 17 \\ +\ 2 \\ \hline \end{array}$	$\begin{array}{r} 19 \\ -\ 2 \\ \hline \end{array}$	$\begin{array}{r} 12 \\ +\ 7 \\ \hline \end{array}$	$\begin{array}{r} 19 \\ -\ 7 \\ \hline \end{array}$
# 4	$\begin{array}{r} 16 \\ +\ 3 \\ \hline \end{array}$	$\begin{array}{r} 19 \\ -\ 3 \\ \hline \end{array}$	$\begin{array}{r} 11 \\ +\ 8 \\ \hline \end{array}$	$\begin{array}{r} 19 \\ -\ 8 \\ \hline \end{array}$
# 5	$\begin{array}{r} 15 \\ +\ 4 \\ \hline \end{array}$	$\begin{array}{r} 19 \\ -\ 4 \\ \hline \end{array}$	$\begin{array}{r} 10 \\ +\ 9 \\ \hline \end{array}$	$\begin{array}{r} 19 \\ -\ 9 \\ \hline \end{array}$
# 6	$\begin{array}{r} 9 \\ +\ 10 \\ \hline \end{array}$	$\begin{array}{r} 19 \\ -\ 10 \\ \hline \end{array}$	$\begin{array}{r} 4 \\ +\ 15 \\ \hline \end{array}$	$\begin{array}{r} 19 \\ -\ 15 \\ \hline \end{array}$
# 7	$\begin{array}{r} 8 \\ +\ 11 \\ \hline \end{array}$	$\begin{array}{r} 19 \\ -\ 11 \\ \hline \end{array}$	$\begin{array}{r} 3 \\ +\ 16 \\ \hline \end{array}$	$\begin{array}{r} 19 \\ -\ 16 \\ \hline \end{array}$
# 8	$\begin{array}{r} 7 \\ +\ 12 \\ \hline \end{array}$	$\begin{array}{r} 19 \\ -\ 12 \\ \hline \end{array}$	$\begin{array}{r} 2 \\ +\ 17 \\ \hline \end{array}$	$\begin{array}{r} 19 \\ -\ 17 \\ \hline \end{array}$
# 9	$\begin{array}{r} 6 \\ +\ 13 \\ \hline \end{array}$	$\begin{array}{r} 19 \\ -\ 13 \\ \hline \end{array}$	$\begin{array}{r} 1 \\ +\ 18 \\ \hline \end{array}$	$\begin{array}{r} 19 \\ -\ 18 \\ \hline \end{array}$
# 10	$\begin{array}{r} 5 \\ +\ 14 \\ \hline \end{array}$	$\begin{array}{r} 19 \\ -\ 14 \\ \hline \end{array}$	$\begin{array}{r} 0 \\ +\ 19 \\ \hline \end{array}$	$\begin{array}{r} 19 \\ -\ 19 \\ \hline \end{array}$

Summary 1 Chapter 2 Lesson 16
Test 2 A
Add or Subtract numbers in row

# 1	9 + 5	# 6	15 – 6
# 2	6 + 9	# 7	15 – 8
# 3	9 + 7	# 8	16 – 7
# 4	8 + 9	# 9	17 – 8
# 5	9 + 9	# 10	16 – 9

Add or Subtract numbers in column

# 11	9 + 7	# 12	6 + 9	# 13	8 + 7	# 14	7 + 8	# 15	9 + 6
# 16	14 -5	# 17	14 - 7	# 18	14 - 9	# 19	13 - 6	# 20	13 - 8

Summary 1 Chapter 2 Lesson 16
Test 2 B
Add or Subtract numbers in row

1 5 + 9

2 9 + 6

3 7 + 9

4 9 + 8

5 9 + 9

6 15 – 7

7 15 – 9

8 16 – 8

9 17 – 9

10 18 – 9

Add or Subtract numbers in column

# 11 8 + 5	# 12 7 + 6	# 13 6 + 7	# 14 5 + 8	# 15 4 + 9
# 16 14 -6	# 17 14 - 8	# 18 13 - 5	# 19 13 - 7	# 20 13 - 9

Summary 1 Answers Chapter 2

Lesson 1

# 1	9	# 2	6	# 3	7	# 4	8	# 5	9
# 6	6	# 7	2	# 8	6	# 9	2	# 10	7
# 11	8	# 12	9	# 13	9	# 14	9	# 15	9
# 16	4	# 17	0	# 18	4	# 19	0	# 20	6

Lesson 3

	a	b	c	d	e
# 1	<	>	<	>	<
# 2	<	>	<	>	>
# 3	<	>	<	<	>
# 4	>	<	>	>	<
# 5	>	>	<	<	<
# 6	>	<	>	<	>
# 7	<	<	>	>	<
# 8	>	<	>	>	<
# 9	>	<	<	>	<
# 10	>	<	>	>	<
# 11	<	>	<	>	<
# 12	>	<	>	>	<
# 13	<	<	>	<	<
# 14	>	<	>	>	>
# 15	<	>	<	<	<
# 16	>	<	>	<	>
# 17	<	>	<	<	>
# 18	>	>	<	>	>
# 19	<	>	<	>	<
# 20	>	<	>	<	>

Summary 1 Answers Chapter 2

Lesson 4

	a	b	c	d	e
# 1	=	=	=	>	>
# 2	<	>	<	<	>
# 3	>	<	>	>	<
# 4	<	>	>	<	<
# 5	>	<	>	>	<
# 6	<	>	<	<	>
# 7	<	>	<	<	>
# 8	>	<	>	>	<
# 9	<	<	>	>	<
# 10	<	>	<	<	>
# 11	=	=	=	>	>
# 12	<	>	<	<	>
# 13	>	<	>	>	<
# 14	<	>	<	<	>
# 15	>	<	<	>	>
# 16	<	>	<	<	>
# 17	=	=	=	>	>
# 18	<	>	<	<	>
# 19	>	<	=	>	<
# 20	<	>	<	<	>

Lesson 5

	a	b	c	d	e
# 1	1	2	3	4	5
# 2	6	7	8	9	10
# 3	2	3	4	5	6
# 4	7	8	9	10	11
# 5	3	4	5	6	7
# 6	8	9	10	11	12
# 7	4	5	6	7	8
# 8	9	10	11	12	13
# 9	5	6	7	8	9
# 10	10	11	12	13	14

Summary 1 Answers Chapter 2

Lesson 5

	a	b	c	d	e
# 11	6	7	8	9	10
# 12	11	12	13	14	15
# 13	7	8	9	10	11
# 14	12	13	14	15	16
# 15	8	9	10	11	12
# 16	13	14	15	16	17
# 17	9	10	11	12	13
# 18	14	15	16	17	18
# 19	10	11	12	13	14
# 20	15	16	17	18	19

Lesson 6

	a	b	c	d	e
# 1	8	9	10	11	12
# 2	13	14	15	16	17
# 3	9	10	11	12	13
# 4	14	15	16	17	18
# 5	10	11	12	13	14
# 6	15	16	17	18	19
# 7	11	12	13	14	15
# 8	16	17	18	19	19
# 9	12	13	14	15	16
# 10	17	18	19	19	18
# 11	13	14	15	16	17
# 12	18	19	17	18	19
# 13	14	15	16	17	18
# 14	19	16	17	18	19
# 15	15	16	17	18	19
# 16	15	16	17	18	19
# 17	16	17	18	19	19
# 18	16	17	18	19	19
# 19	17	18	19	18	19
# 20	17	18	19	18	19

Summary 1 Answers Chapter 2

Lesson 7

	a	b	c	d	e
# 1	10	9	8	7	6
# 2	5	4	3	2	1
# 3	11	10	9	8	7
# 4	6	5	4	3	2
# 5	12	11	10	9	8
# 6	7	6	5	4	3
# 7	13	12	11	10	9
# 8	8	7	6	5	4
# 9	14	13	12	11	10
# 10	9	8	7	6	5
# 11	15	14	13	12	11
# 12	10	9	8	7	6
# 13	16	15	14	13	12
# 14	11	10	9	8	7
# 15	17	16	15	14	13
# 16	12	11	10	9	8
# 17	18	17	16	15	14
# 18	13	12	11	10	9
# 19	19	18	17	16	15
# 20	14	13	12	11	10

Lesson 8

	a	b	c	d
# 1	10	9	10	4
# 2	10	8	10	3
# 3	10	7	10	2
# 4	10	6	10	1
# 5	10	5	10	0
# 6	11	10	11	5
# 7	11	9	11	4
# 8	11	8	11	3
# 9	11	7	11	2
# 10	11	6	11	1

Summary 1 Answers Chapter 2

Lesson 9

	a	b	c	d
# 1	12	12	12	7
# 2	12	11	12	6
# 3	12	10	12	5
# 4	12	9	12	4
# 5	12	8	12	3
# 6	13	13	13	8
# 7	13	12	13	7
# 8	13	11	13	6
# 9	13	10	13	5
# 10	13	9	13	4

Lesson 10

	a	b	c	d
# 1	14	14	14	9
# 2	14	13	14	8
# 3	14	12	14	7
# 4	14	11	14	6
# 5	14	10	14	5
# 6	14	4	12	12
# 7	14	3	12	12
# 8	14	2	12	3
# 9	14	1	12	2
# 10	14	0	0	1

Lesson 11

	a	b	c	d
# 1	15	15	15	10
# 2	15	14	15	9
# 3	15	13	15	8
# 4	15	12	15	7
# 5	15	11	15	6
# 6	15	5	13	3
# 7	15	4	13	2
# 8	15	3	13	1
# 9	15	2	13	0
# 10	15	1	13	0

Summary 1 Answers Chapter 2

Lesson 12

	a	b	c	d
# 1	16	16	16	11
# 2	16	15	16	10
# 3	16	14	16	9
# 4	16	13	16	8
# 5	16	12	16	7
# 6	16	6	16	11
# 7	16	5	16	10
# 8	16	4	16	9
# 9	16	3	16	8
# 10	16	2	16	7

Lesson 13

	a	b	c	d
# 1	17	17	17	12
# 2	17	16	17	11
# 3	17	15	17	10
# 4	17	14	17	9
# 5	17	13	17	8
# 6	17	7	17	12
# 7	17	6	17	11
# 8	17	5	17	10
# 9	17	4	17	9
# 10	17	3	17	8

Lesson 14

	a	b	c	d
# 1	18	18	18	13
# 2	18	17	18	12
# 3	18	16	18	11
# 4	18	15	18	10
# 5	18	14	18	9
# 6	18	8	18	13
# 7	18	7	18	12
# 8	18	6	18	11
# 9	18	5	18	10
# 10	18	4	18	9

Summary 1 Answers Chapter 2

Lesson 15

	a	b	c	d
# 1	19	19	19	14
# 2	19	18	19	13
# 3	19	17	19	12
# 4	19	16	19	11
# 5	19	15	19	10
# 6	19	9	19	4
# 7	19	8	19	3
# 8	19	7	19	2
# 9	19	6	19	1
# 10	19	5	19	0

Lesson 16

Test 2 A

# 1	14	# 2	15	# 3	16	# 4	17	# 5	18
# 6	9	# 7	7	# 8	9	# 9	9	# 10	7
# 11	16	# 12	15	# 13	15	# 14	15	# 15	15
# 16	9	# 17	7	# 18	5	# 19	7	# 20	5

Test 2 B

# 1	14	# 2	15	# 3	16	# 4	17	# 5	18
# 6	8	# 7	6	# 8	8	# 9	8	# 10	9
# 11	13	# 12	13	# 13	13	# 14	13	# 15	13
# 16	8	# 17	6	# 18	8	# 19	6	# 20	4

Summary 1

Chapter 3

Three

Numbers

Summary 1 Chapter 3 Contents
Three Numbers
Lesson 1 - 14
Basic Problems

Work with lesson 1, read rules and examples, 2) Solve the problems from contents, 3) Check answers in the end of chapter 1, 4) Work with the next lessons 2, 3, 4, 5, ..., 5) Solve Test 1 A, 6) Grading Scale: A =10 points, B=9points, C=8 points, D=7 points, F=0-6 points, 7) If you solve 10 problems correctly, you get an A, 8) If have an A, you can move on to next chapter, 9) If your grade is lower than an A, work with the mistakes, then solve Test 1B, 10) When you finish all of the chapters, solve Final Exam in last chapter.   Lesson = L

Summary 1 Chapter 3 Lesson 1
Three Numbers
Basic Problems
Addition three numbers in row 1
<u>Three numbers can add in any order</u>

$$9 + 4 + 3 = 13 + 3 = 16$$
$$9 + 4 + 3 = 9 + 7 = 16$$
$$9+3+4=12+4=16$$
$$9+3+4=9+7=16$$

Add numbers

	a	b	c	d
# 1	0 + 1 + 0	0 + 2 + 0	0 + 3 + 0	0 + 4 + 0
# 2	1 + 1 + 1	1 + 2 + 1	1 + 3 + 1	1 + 4 + 1
# 3	2 + 1 + 2	2 + 2 + 2	2 + 3 + 2	2 + 4 + 2
# 4	3 + 1 + 3	3 + 2 + 3	3 + 3 + 3	3 + 4 + 3
# 5	4 + 1 + 4	4 + 2 + 4	4 + 3 + 4	4 + 4 + 4
# 6	5 + 1 + 5	5 + 2 + 5	5 + 3 + 5	5 + 4 + 5
# 7	6 + 1 + 6	6 + 2 + 6	6 + 3 + 6	6 + 4 + 6
# 8	7 + 1 + 7	7 + 2 + 7	7 + 3 + 7	7 + 4 + 7
# 9	8 + 1 + 8	8 + 2 + 8	8 + 3 + 8	2 + 9 + 2
# 10	9 + 1 + 9	0 + 9 + 0	1 + 9 + 1	3 + 9 + 3
# 11	0 + 5 + 0	0 + 6 + 0	0 + 7 + 0	0 + 8 + 0
# 12	1 + 5 + 1	1 + 6 + 1	1 + 7 + 1	1 + 8 + 1
# 13	2 + 5 + 2	2 + 6 + 2	2 + 7 + 2	2 + 8 + 2
# 14	3 + 5 + 3	3 + 6 + 3	3 + 7 + 3	3 + 8 + 3
# 15	4 + 5 + 4	4 + 6 + 4	4 + 7 + 4	4 + 8 + 4
# 16	5 + 5 + 5	5 + 6 + 5	5 + 7 + 5	5 + 8 + 5
# 17	6 + 5 + 6	6 + 6 + 6	6 + 7 + 6	6 + 1 + 2
# 18	7 + 5 + 7	7 + 1 + 2	3 + 1 + 2	7 + 1 + 2
# 19	4 + 9 + 4	1 + 1 + 2	4 + 1 + 2	8 + 1 + 2
# 20	5 + 9 + 5	2 + 1 + 2	5 + 1 + 2	9 + 1 + 2

Summary 1 Chapter 3 Lesson 2

Addition three numbers in row 2

<u>Three numbers can add in any order</u>

9 + 5 + 3 = 14 + 3 = 17

9 + 5 + 3 = 9 + 8 = 17

9+3+5=12+5=17

9+3+5 = 9+8=17

Add numbers

	a	b	c	d
# 1	0 + 1 + 2	0 + 2 + 3	0 + 3 + 4	0 + 4 + 5
# 2	1 + 1 + 2	1 + 2 + 3	1 + 3 + 4	1 + 4 + 5
# 3	2 + 1 + 2	2 + 2 + 3	2 + 3 + 4	2 + 4 + 5
# 4	3 + 1 + 2	3 + 2 + 3	3 + 3 + 4	3 + 4 + 5
# 5	4 + 1 + 2	4 + 2 + 3	4 + 3 + 4	4 + 4 + 5
# 6	5 + 1 + 2	5 + 2 + 3	5 + 3 + 4	5 + 4 + 5
# 7	6 + 1 + 2	6 + 2 + 3	6 + 3 + 4	6 + 4 + 5
# 8	7 + 1 + 2	7 + 2 + 3	7 + 3 + 4	7 + 4 + 5
# 9	8 + 1 + 2	8 + 2 + 3	8 + 3 + 4	8 + 4 + 5
# 10	9 + 1 + 2	9 + 2 + 3	9 + 3 + 4	9 + 4 + 5
# 11	0 + 1 + 3	0 + 2 + 4	0 + 3 + 4	0 + 4 + 5
# 12	1 + 1 + 3	1 + 2 + 4	1 + 3 + 4	1 + 4 + 5
# 13	2 + 1 + 3	2 + 2 + 4	2 + 3 + 4	2 + 4 + 5
# 14	3 + 1 + 3	3 + 2 + 4	3 + 3 + 4	3 + 4 + 5
# 15	4 + 1 + 3	4 + 2 + 4	4 + 3 + 4	4 + 4 + 5
# 16	5 + 1 + 3	5 + 2 + 4	5 + 3 + 4	5 + 4 + 5
# 17	6 + 1 + 3	6 + 2 + 4	6 + 3 + 4	6 + 4 + 5
# 18	7 + 1 + 3	7 + 2 + 4	7 + 3 + 4	7 + 4 + 5
# 19	8 + 1 + 3	8 + 2 + 4	8 + 3 + 4	8 + 4 + 5
# 20	9 + 1 + 3	9 + 2 + 4	9 + 3 + 4	9 + 4 + 5

Summary 1 Chapter 3 Lesson 3
Subtraction three numbers in row 1
<u>To subtract numbers, work from left to right</u>

10 – 1 – 4 = 9 – 4 = 5
10 – 1 – 5 = 9 – 5 = 4

Subtract numbers

	a	b	c	d
# 1	1 - 1 - 0	1 - 0 - 1	2 - 1 - 1	4 – 1 - 2
# 2	2 - 1 - 0	2 - 0 - 1	3 - 1 – 1	5 – 1 - 2
# 3	3 - 1 - 0	3 - 0 - 1	4 - 1 – 1	6 – 1 - 2
# 4	4 - 1 - 0	4 - 0 - 1	5 - 1 – 1	7 – 1 - 2
# 5	5 - 1 - 0	5 - 0 - 1	6 - 1 – 1	8 – 1 - 2
# 6	6 - 1 - 0	6 - 0 - 1	7 - 1 – 1	9 – 1 - 2
# 7	7 - 1 - 0	7 - 0 - 1	8 - 1 – 1	10 – 1 - 2
# 8	8 - 1 - 0	8 - 0 - 1	9 - 1 – 1	11 – 1 - 2
# 9	9 - 1 - 0	9 - 0 - 1	10 - 1 – 1	12 – 1 - 2
# 10	10 - 1 - 0	10 - 0 - 1	11 - 1 – 1	13 – 1 - 2
# 11	4 - 2 - 1	5 - 1 - 3	5 - 3 - 1	5 – 2 - 2
# 12	5 - 2 - 1	6 - 1 - 3	6 - 3 – 1	6 – 2 - 2
# 13	6 - 2 - 1	7 - 1 - 3	7 - 3 - 1	7 – 2 - 2
# 14	7 - 2 - 1	8 - 1 - 3	8 - 3 – 1	8 – 2 - 2
# 15	8 - 2 - 1	9 - 1 - 3	9 - 3 - 1	9 – 2 - 2
# 16	9 - 2 - 1	10 - 1 - 3	10 - 3 – 1	10 – 2 - 2
# 17	10 - 2 - 1	11 - 1 - 3	11 - 3 – 1	11 – 2 - 2
# 18	11 - 2 - 1	12 - 1 - 3	12 - 3 – 1	12 – 2 - 2
# 19	12 - 2 - 1	13 - 1 - 3	13 - 3 – 1	13 – 2 - 2
# 20	13 - 2 - 1	14 - 1 - 3	14 - 3 – 1	14 – 2 - 2

Summary 1 Chapter 3 Lesson 4

Subtraction three numbers in row 2

<u>To subtract numbers, work from left to right</u>

10 – 1 – 6 = 9 – 6 = 3

10 – 1 – 7 = 9 – 7 = 2

Subtract numbers

	a	b	c	d
# 1	6 - 1 - 4	6 - 4 - 1	6 - 2 - 3	6 – 3 - 2
# 2	7 - 1 - 4	7 - 4 - 1	7 - 2 – 3	7 – 3 - 2
# 3	8 - 1 - 4	8 - 4 - 1	8 - 2 – 3	8 – 3 - 2
# 4	9 - 1 - 4	9 - 4 - 1	9 - 2 – 3	9 – 3 - 2
# 5	10 - 1 - 4	10 - 4 - 1	10 - 2 – 3	10 – 3 - 2
# 6	11 - 1 - 4	11 - 4 - 1	11 - 2 – 3	11 – 3 - 2
# 7	12 - 1 - 4	12 - 4 - 1	12 - 2 – 3	12 – 3 - 2
# 8	13 - 1 - 4	13 - 4 - 1	13 - 2 – 3	13 – 3 – 2
# 9	14 - 1 - 4	14 - 4 - 1	14 - 2 – 3	14 – 3 - 2
# 10	15 - 1 - 4	15 - 4 - 1	15 - 2 – 3	15 – 3 - 2
# 11	7 - 1 - 6	7 - 6 - 1	7 - 2 - 5	7 – 5 - 2
# 12	8 - 1 - 6	8 - 6 - 1	8 - 2 – 5	8 – 5 - 2
# 13	9 - 1 - 6	9 - 6 - 1	9 - 2 - 5	9 – 5 - 2
# 14	10 - 1 - 6	10 - 6 - 1	10 - 2– 5	10 – 5 - 2
# 15	11 - 1 - 6	11 - 6 - 1	11 - 2 - 5	11 – 5 - 2
# 16	12 - 1 - 6	12 - 6 – 1	12 - 2 – 5	12 – 5 - 2
# 17	13 - 1 - 6	13 - 6 - 1	13 - 2 – 5	13 – 5 - 2
# 18	14 - 1 - 6	14 - 6 - 1	14 - 2 – 5	14 – 5 - 2
# 19	15 - 1 - 6	15 - 6 - 1	15 - 2 – 5	15 – 5 - 2
# 20	16 - 1 - 6	16 - 6 - 1	16 - 2 – 5	16 – 5 - 2

Summary 1 Chapter 3 Lesson 5

Remember

10 = 9 + 1= 8 + 2= 7 + 3 = 6 + 4 = 5 + 5

11 = 9 + 2 = 8 + 3 = 7 + 4 = 6 + 5

12 = 9 +3 = 8 + 4 = 7 + 5 = 6 + 6   13 = 9 + 4 = 8 + 5 = 7 + 6

Two operations with three numbers in row 1

<u>To add and subtract numbers, work from left to right</u>

5 + 6 – 9 = 11 – 9 = 2   5 + 7 – 9 = 12 - 9 = 3

Add and Subtract numbers

	a	b	c	d
# 1	1 + 8 - 0	2 + 7 - 0	4 + 6 – 0	5 + 9 - 0
# 2	1 + 8 - 1	2 + 7 - 1	4 + 6 – 1	5 + 9 - 1
# 3	1 + 8 - 2	2 + 7 - 2	4 + 6 – 2	5 + 9 - 2
# 4	1 + 8 - 3	2 + 7 - 3	4 + 6 – 3	5 + 9 - 3
# 5	1 + 8 - 4	2 + 7 - 4	4 + 6 – 4	5 + 9 - 4
# 6	1 + 8 - 5	2 + 7 - 5	4 + 6 – 5	5 + 9 - 5
# 7	1 + 8 - 6	2 + 7 - 6	4 + 6 – 6	5 + 9 - 6
# 8	1 + 8 - 7	2 + 7 - 7	4 + 6 – 7	5 + 9 - 7
# 9	1 + 8 - 8	2 + 7 - 8	4 + 6 – 8	5 + 9 - 8
# 10	1 + 8 - 9	2 + 7 - 9	4 + 6 – 9	5 + 9 - 9
# 11	6 + 4 - 0	3 + 7 - 0	7 + 4 - 0	8 + 5 - 0
# 12	6 + 4 - 1	3 + 7 - 1	7 + 4 – 1	8 + 5 - 1
# 13	6 + 4 - 2	3 + 7 - 2	7 + 4 - 2	8 + 5 - 2
# 14	6 + 4 - 3	3 + 7 - 3	7 + 4 – 3	8 + 5 - 3
# 15	6 + 4 - 4	3 + 7 - 4	7 + 4 - 4	8 + 5 - 4
# 16	6 + 4 - 5	3 + 7 – 5	7 + 4 – 5	8 + 5 - 5
# 17	6 + 4 - 6	3 + 7 - 6	7 + 4 – 6	8 + 5 - 6
# 18	6 + 4 - 7	3 + 7 - 7	7 + 4 – 7	8 + 5 - 7
# 19	6 + 4 - 8	3 + 7 - 8	7 + 4 – 8	8 + 5 - 8
# 20	6 + 4 - 9	3 + 7 - 9	7 + 4 – 9	8 + 5 - 9

Summary 1 Chapter 3 Lesson 6

Remember

14 = 9 + 5 = 8 + 6 = 7 + 7	15 = 9 + 6 = 8 + 7
16 = 9 + 7 = 8 + 8	17 = 9 + 8
18 = 9 + 9	19 = 10 + 9

Two operations with three numbers in row 2

<u>To subtract and add numbers, work from left to right</u>

9 - 8 + 7 = 1 + 7 = 8

8 – 7 + 6 = 1 + 6 = 7

Subtract and add numbers

	a	b	c	d
# 1	4 - 3 + 0	5 - 3 + 0	6 - 3 + 0	7 – 3 + 0
# 2	4 - 3 + 1	5 - 3 + 1	6 - 3 + 1	7 – 3 + 1
# 3	4 - 3 + 2	5 - 3 + 2	6 - 3 + 2	7 – 3 + 2
# 4	4 - 3 + 3	5 - 3 + 3	6 - 3 + 3	7 – 3 + 3
# 5	4 - 3 + 4	5 - 3 + 4	6 - 3 + 4	7 – 3 + 4
# 6	4 - 3 + 5	5 - 3 + 5	6 - 3 + 5	7 – 3 + 5
# 7	4 - 3 + 6	5 - 3 + 6	6 - 3 + 6	7 – 3 + 6
# 8	4 - 3 + 7	5 - 3 + 7	6 - 3 + 7	7 – 3 + 7
# 9	4 - 3 + 8	5 - 3 + 8	6 - 3 + 8	7 – 3 + 8
# 10	4 - 3 + 9	5 - 3 + 9	6 - 3 + 9	7 – 3 + 9
# 11	6 - 4 + 0	7 - 4 + 0	8 - 4 + 0	9 – 4 + 0
# 12	6 - 4 + 1	7 - 4 + 1	8 - 4 + 1	9 – 4 + 1
# 13	6 - 4 + 2	7 - 4 + 2	8 - 4 + 2	9 – 4 + 2
# 14	6 - 4 + 3	7 - 4 + 3	8 – 4 + 3	9 – 4 + 3
# 15	6 - 4 + 4	7 - 4 + 4	8 - 4 + 4	9 – 4 + 4
# 16	6 - 4 + 5	7 - 4 + 5	8 - 4 + 5	9 – 4 + 5
# 17	6 - 4 + 6	7 - 4 + 6	8 - 4 + 6	9 – 4 + 6
# 18	6 - 4 + 7	7 - 4 + 7	8 - 4 + 7	9 – 4 + 7
# 19	6 - 4 + 8	7 - 4 + 8	8 - 4 + 8	9 – 4 + 8
# 20	6 - 4 + 9	7 - 4 + 9	8 - 4 + 9	9 – 4 + 9

Summary 1 Chapter 3 Lesson 7
Addition tens in column

80 + 10 = 90 or 10 + 80 = 90

80	1) 0 + 0 = 0	10	1) 0 + 0 =0
+ 10	2) Put 0 below 0	+80	2) Put 0 below 0
90	3) 8 + 1 = 9	90	3) 1 + 8 = 9
	4) Put 9 below 1		4) Put 9 below 8

Add tens

	a	b	c	d
# 1	10 + 10	10 + 20	10 + 30	10 + 40
# 2	10 + 50	10 + 60	10 + 70	10 + 80
# 3	20 + 10	20 + 20	20 + 30	20 + 40
# 4	20 + 50	20 + 60	20 + 70	70 + 20
# 5	30 +10	30 + 20	30 + 30	30 + 40
# 6	30 + 50	30 + 60	50 + 30	60 + 30
# 7	40 + 10	40 + 20	40 + 30	40 + 40
# 8	40 + 50	20 + 40	30 + 40	20 + 60
# 9	50 + 10	50 + 20	50 + 30	50 + 40
# 10	60 + 10	60 + 20	60 + 30	30 + 60

Summary 1 Chapter 3 Lesson 8
Subtraction tens in column

90 – 10 = 80 or 90 – 80 = 10

90	1) 0 - 0 = 0	90	1) 0 – 0 = 0
- 10	2) Put 0 below 0	- 80	2) Put 0 below 0
80	3) 9 – 1 = 8	10	3) 9 – 8 = 1
	4) Put 8 below 1		4) Put 1 below 8

Subtract tens

	a	b	c	d
# 1	90 - 10	90 - 20	90 - 30	90 - 40
# 2	90 - 50	90 - 60	90 - 70	90 - 80
# 3	90 - 90	80 - 10	80 - 20	80 - 30
# 4	80 - 40	80 - 50	80 - 60	80 - 70
# 5	80 - 80	70 - 10	70 - 20	70 - 30
# 6	70 - 40	70 - 50	70 - 60	70 - 70
# 7	60 - 10	60 - 20	60 - 30	60 - 40
# 8	60 - 50	60 - 60	50 - 10	50 - 20
# 9	50 - 30	50 - 40	50 - 50	40 - 10
# 10	40 - 20	40 - 30	30 - 10	30 - 20

Summary 1 Chapter 3 Lesson 9

Addition two-digit numbers in column

Add the ones and then add the tens

43	1) 3 + 5 =8	27	1) 7 + 1 = 8
+ 25	2) Put 8 below 5	+ 51	2) Put 8 below 1
68	3) 4 + 2 = 6	78	3) 2 + 5 = 7
	4) Put 6 below 2		4) Put 7 below 5

Add numbers

	a	b	c	d
# 1	20 + 79	21 + 68	22 + 57	23 + 46
# 2	24 + 35	25 + 23	26 + 12	27 + 41
# 3	28 + 70	27 + 60	30 + 58	31 + 66
# 4	32 + 65	33 + 53	34 + 42	35 + 34
# 5	36 + 20	37 + 10	38 + 31	39 + 10
# 6	40 + 25	41 + 35	42 + 44	43 + 52
# 7	44 + 40	45 + 30	46 + 21	47 + 11
# 8	48 + 20	49 + 30	50 + 42	51 + 36
# 9	52 + 45	53 + 30	54 + 21	55 + 11
# 10	56 + 23	50 + 32	58 + 41	59 + 30

Summary 1 Chapter 3 Lesson 10

Subtraction two-digit numbers in column

Subtract the ones then and subtract the tens

96	1) 6 – 5 = 1	89	1) 9 – 2 = 7
- 25	2) Put 1 below 5	- 52	2) Put 7 below 2
71	3) 9 – 2 = 7	37	3) 8 – 5 = 3
	4) Put 7 below 2		4) Put 3 below 5

Subtract numbers

	a	b	c	d
# 1	99 - 87	98 - 76	97 - 65	96 - 54
# 2	95 - 43	94 - 32	93 - 21	92 - 10
# 3	91 - 20	90 - 30	89 - 55	88 - 74
# 4	87 - 56	86 - 43	85 - 41	84 - 31
# 5	85 - 31	82 - 20	81 - 11	86 - 14
# 6	79 - 22	78 - 33	77 - 44	76 - 50
# 7	75 - 61	74 - 52	73 - 41	72 - 31
# 8	79 - 44	70 - 40	69 - 59	68 - 43
# 9	67 - 54	66 - 41	65 - 32	64 - 21
# 10	97 - 52	86 - 42	75 - 34	69 - 35

Summary 1 Chapter 3 Lesson 11
Addition three numbers in column
Add the ones and then add the tens

10	1) 0 + 1 + 2 = 3
+ 21	2) Put 3 below 2
32	3) 1 + 2 + 3 = 6
63	4) Put 6 below 3

Add numbers

	a	b	c	d
# 1	20 + 12 32	21 + 13 31	22 + 14 30	23 + 15 30
# 2	24 + 11 23	25 + 12 22	26 + 11 20	23 + 10 22
# 3	28 + 20 21	20 + 21 22	30 + 22 23	31 + 23 22
# 4	32 + 21 20	33 + 25 10	34 + 23 12	33 + 20 14
# 5	31 + 22 12	32 + 24 11	33 + 22 11	34 + 21 21

Summary 1 Chapter 3 Lesson 11
Addition three numbers in column
Add the ones and then add the tens

21	1) 1 + 2 + 3 = 6
+ 32	2) Put 6 below 3
43	3) 2 + 3 + 4 = 9
96	4) Put 9 below 4

Add numbers

	a	b	c	d
# 6	36 + 20 12	37 + 20 12	36 + 20 11	30 + 31 13
# 7	40 + 32 12	41 + 33 11	42 + 34 10	43 + 35 11
# 8	44 + 32 12	45 + 30 11	46 + 32 11	44 + 31 10
# 9	48 + 40 11	43 + 40 10	50 + 32 12	31 + 43 12
# 10	21 + 31 24	34 + 22 21	24 + 32 22	35 + 21 23

Summary 1 Chapter 3 Lesson 12
Add and Subtract numbers
Test 3 A
Add and Subtract numbers in row

 # 1 9 + 6 + 4
 # 2 8 + 6 + 5
 # 3 10 – 3 - 1
 # 4 10 – 4 - 2
 # 5 6 + 6 - 2

 # 6 6 + 7 - 3
 # 7 10 - 4 + 3
 # 8 10 – 5 + 4
 # 9 60 + 20
 # 10 60 - 20

Add and Subtract numbers in column

# 11	# 12	# 13	# 14	# 15
10	90	51	95	35
+ 80	- 70	+ 48	- 41	+ 23
				11

# 16	# 17	# 18	# 19	# 20
20	80	62	84	24
+ 60	- 50	+ 36	- 31	+ 21
				32

Summary 1 Chapter 3 Lesson 12
Add and Subtract numbers
Test 3 B
Add and Subtract Numbers in row

 # 1 7 + 5 + 4
 # 2 6 + 4 + 3
 # 3 10 – 5 - 3
 # 4 10 – 6 – 4
 # 5 6 + 8 - 4

 # 6 6 + 9 - 5
 # 7 10 – 6 + 5
 # 8 10 – 7 + 6
 # 9 40 + 20
 # 10 40 – 20

Add and Subtract Numbers in column

# 11	# 12	# 13	# 14	# 15
30	70	73	74	42
+ 40	- 30	+ 24	- 21	+ 20
				34

# 16	# 17	# 18	# 19	# 20
60	60	84	62	51
+ 20	- 10	+12	- 11	+ 13
				23

Summary 1 Chapter 3 Lesson 13
Review 1 A
1-20 odd or # 1, 3, 5, 7, 9, 11, 13, 15, 17, 19

1 $5 + 0 + 1$

2 $6 + 1 + 2$

3 $7 + 2 + 3$

4 $8 + 3 + 4$

5 $9 + 4 + 5$

6 $15 - 0 - 1$

7 $16 - 1 - 2$

8 $17 – 2 - 3$

9 $18 - 3 - 4$

10 $19 – 4 - 5$

Add and Subtract numbers in column

# 11	# 12	# 13	# 14	# 15
20	42	35	65	34
+ 30	+ 25	+ 43	+ 21	+ 43
				21

# 16	# 17	# 18	# 19	# 20
80	98	48	37	54
- 30	-35	- 23	- 14	+ 13
				12

Summary 1 Chapter 3 Lesson 13
Review 1 B
Add and Subtract numbers in row
1-20 odd or # 1, 3, 5, 7, 9, 11, 13, 15, 17, 19

# 1	5 + 1 + 2
# 2	6 + 2 + 3
# 3	7 + 4 + 4
# 4	8 + 4 + 5
# 5	9 + 0 + 1
# 6	15 - 1 - 2
# 7	16 - 2 - 3
# 8	17 – 3 - 4
# 9	18 - 4 - 5
# 10	19 – 0 - 1

Add and Subtract numbers in column

# 11	40 + 20	# 12	53 + 42	# 13	35 + 14	# 14	25 + 32	# 15	14 + 53 11
# 16	60 - 40	# 17	94 -23	# 18	47 -24	# 19	57 - 16	# 20	64 + 23 12

Summary 1 Chapter 3 Lesson 14
Final Exam 1 A
Add and Subtract numbers in row
1-20 odd or # 1, 3, 5, 7, 9, 11, 13, 15, 17, 19

1 9 + 5 + 4

2 8 + 4 + 3

3 7 + 3 + 2

4 6 + 2 + 1

5 5 + 1 + 0

6 19 - 5 - 4

7 18 - 4 - 3

8 17 – 3 - 2

9 16 – 2 - 1

10 15 – 1 - 0

Add and Subtract numbers in column

# 11	# 12	# 13	# 14	# 15
20	80	15	59	53
+ 70	- 50	+ 84	- 14	+ 32
				11

# 16	# 17	# 18	# 19	# 20
30	80	26	48	42
+ 50	- 60	+ 63	- 15	+ 12
				23

Summary 1 Chapter 3 Lesson 14
Final Exam 1 B
Add and Subtract Numbers in row
1-20 odd or # 1, 3, 5, 7, 9, 11, 13, 15, 17, 19

1   9 + 1 + 0

2   8 + 1 + 2

3   7 + 2 + 3

4   6 + 3 + 4

5   5 + 4 + 5

6   19 - 1 - 0

7   18 – 2 - 1

8   17 – 3 - 2

9   16 – 4 - 3

10   15 – 5 - 4

Add and Subtract Numbers in column

# 11	# 12	# 13	# 14	# 15
40	60	37	47	24
+ 20	- 10	+ 42	- 12	+ 12
				43

# 16	# 17	# 18	# 19	# 20
20	50	48	26	15
+ 70	- 10	+21	- 14	+ 31
				32

Summary 1 Answers Chapter 3

Lesson 1

	a	b	c	d
# 1	1	2	3	4
# 2	3	4	5	6
# 3	5	6	7	8
# 4	7	8	9	10
# 5	9	10	11	12
# 6	11	12	13	14
# 7	13	14	15	16
# 8	15	16	17	18
# 9	17	18	19	13
# 10	19	9	11	15

Lesson 2

	a	b	c	d
# 1	3	5	7	9
# 2	4	6	8	10
# 3	5	7	9	11
# 4	6	8	10	12
# 5	7	9	11	13
# 6	8	10	12	14
# 7	9	11	13	15
# 8	10	12	14	16
# 9	11	13	15	17
# 10	12	14	16	18

Lesson 3

	a	b	c	d
# 1	0	0	0	1
# 2	1	1	1	2
# 3	2	2	2	3
# 4	3	3	3	4
# 5	4	4	4	5
# 6	5	5	5	6
# 7	6	6	6	7
# 8	7	7	7	8
# 9	8	8	8	9
# 10	9	9	9	10

Lesson 1

	a	b	c	d
# 11	5	6	7	8
# 12	7	8	9	10
# 13	9	10	11	12
# 14	11	12	13	14
# 15	13	14	15	16
# 16	15	16	17	18
# 17	17	18	19	9
# 18	19	10	6	10
# 19	17	4	7	11
# 20	19	5	8	12

Lesson 2

	a	b	c	d
# 11	4	6	7	9
# 12	5	7	8	10
# 13	6	8	9	11
# 14	7	9	10	12
# 15	8	10	11	13
# 16	9	11	12	14
# 17	10	12	13	15
# 18	11	13	14	16
# 19	12	14	15	17
# 20	13	15	16	18

Summary 1 Answers Chapter 3

Lesson 3

	a	b	c	d
# 11	1	1	1	1
# 12	2	2	2	2
# 13	3	3	3	3
# 14	4	4	4	4
# 15	5	5	5	5
# 16	6	6	6	6
# 17	7	7	7	7
# 18	8	8	8	8
# 19	9	9	9	9
# 20	10	10	10	10

Lesson 4

	a	b	c	d
# 1	1	1	1	1
# 2	2	2	2	2
# 3	3	3	3	3
# 4	4	4	4	4
# 5	5	5	5	5
# 6	6	6	6	6
# 7	7	7	7	7
# 8	8	8	8	8
# 9	9	9	9	9
# 10	10	10	10	10

Lesson 4

	a	b	c	d
# 11	0	0	0	0
# 12	1	1	1	1
# 13	2	2	2	2
# 14	3	3	3	3
# 15	4	4	4	4
# 16	5	5	5	5
# 17	6	6	6	6
# 18	7	7	7	7
# 19	8	8	8	8
# 20	9	9	9	9

Summary 1 Answers Chapter 3

Lesson 5

	a	b	c	d
# 1	9	9	10	14
# 2	8	8	9	13
# 3	7	7	8	12
# 4	6	6	7	11
# 5	5	5	6	10
# 6	4	4	5	9
# 7	3	3	4	8
# 8	2	2	3	7
# 9	1	1	2	6
# 10	0	0	1	5

Lesson 5

	a	b	c	d
# 11	10	10	11	13
# 12	9	9	10	12
# 13	8	8	9	11
# 14	7	7	8	10
# 15	6	6	7	9
# 16	5	5	6	8
# 17	4	4	5	7
# 18	3	3	4	6
# 19	2	2	3	5
# 20	1	1	2	4

Lesson 6

	a	b	c	d
# 1	1	2	3	4
# 2	2	3	4	5
# 3	3	4	5	6
# 4	4	5	6	7
# 5	5	6	7	8
# 6	6	7	8	9
# 7	7	8	9	10
# 8	8	9	10	11
# 9	9	10	11	12
# 10	10	11	12	13

Summary 1 Answers Chapter 3

Lesson 6

	a	b	c	d
# 11	2	3	4	5
# 12	3	4	5	6
# 13	4	5	6	7
# 14	5	6	7	8
# 15	6	7	8	9
# 16	7	8	9	10
# 17	8	9	10	11
# 18	9	10	11	12
# 19	10	11	12	13
# 20	11	12	13	14

Lesson 7

	a	b	c	d
# 1	20	30	40	50
# 2	60	70	80	90
# 3	30	40	50	60
# 4	70	80	90	90
# 5	40	50	60	70
# 6	80	90	80	90
# 7	50	60	70	80
# 8	90	60	70	80
# 9	60	70	80	90
# 10	70	80	90	90

Lesson 8

	a	b	c	d
# 1	80	70	60	50
# 2	40	30	20	10
# 3	0	70	60	50
# 4	40	30	20	10
# 5	0	60	50	40
# 6	30	20	10	0
# 7	50	40	30	20
# 8	10	0	40	30
# 9	20	10	0	30
# 10	20	10	20	10

Summary 1 Answers Chapter 3

Lesson 9

	a	b	c	d
# 1	99	89	79	69
# 2	59	48	38	68
# 3	98	87	88	97
# 4	97	86	76	69
# 5	56	47	69	49
# 6	65	76	86	95
# 7	84	75	67	58
# 8	68	79	92	87
# 9	97	83	75	66
# 10	79	82	99	89

Lesson 10

	a	b	c	d
# 1	12	22	32	42
# 2	52	62	72	82
# 3	71	60	34	14
# 4	31	43	44	53
# 5	54	62	70	72
# 6	57	45	33	26
# 7	14	22	32	41
# 8	35	30	10	25
# 9	13	25	33	43
# 10	45	44	41	34

Lesson 11

	a	b	c	d
# 1	64	65	66	68
# 2	58	59	57	55
# 3	69	63	75	76
# 4	73	68	69	67
# 5	65	67	66	76
# 6	68	69	67	74
# 7	84	85	86	89
# 8	88	86	89	85
# 9	99	93	94	86
#10	76	77	78	79

Summary 1 Answers Chapter 3

Lesson 12

Test 3 A

# 1	19	# 2	19	# 3	6	# 4	4	# 5	10
# 6	10	# 7	9	# 8	9	# 9	80	# 10	40
# 11	90	# 12	20	# 13	99	# 14	54	# 15	69
# 16	80	# 17	30	# 18	98	# 19	53	# 20	77

Test 3 B

# 1	16	# 2	13	# 3	2	# 4	0	# 5	10
# 6	10	# 7	9	# 8	9	# 9	60	# 10	20
# 11	70	# 12	40	# 13	97	# 14	53	# 15	96
# 16	80	# 17	50	# 18	96	# 19	51	# 20	87

Lesson 13 Review 1 A

# 1	6	# 2	9	# 3	12	# 4	15	# 5	18
# 6	14	# 7	13	# 8	12	# 9	11	# 10	10
# 11	50	# 12	67	# 13	78	# 14	86	# 15	98
# 16	50	# 17	63	# 18	25	# 19	23	# 20	79

Review 1 B

# 1	8	# 2	11	# 3	15	# 4	17	# 5	10
# 6	12	# 7	11	# 8	10	# 9	9	# 10	18
# 11	60	# 12	95	# 13	49	# 14	57	# 15	78
# 16	20	# 17	71	# 18	23	# 19	41	# 20	99

Lesson 14 Final Exam 1 A

# 1	18	# 2	15	# 3	12	# 4	9	# 5	6
# 6	10	# 7	11	# 8	12	# 9	13	# 10	14
# 11	90	# 12	30	# 13	99	# 14	45	# 15	96
# 16	80	# 17	20	# 18	89	# 19	33	# 20	77

Final Exam 1 B

# 1	10	# 2	11	# 3	12	# 4	13	# 5	14
# 6	18	# 7	15	# 8	12	# 9	9	# 10	6
# 11	60	# 12	50	# 13	79	# 14	35	# 15	79
# 16	90	# 17	40	# 18	69	# 19	12	# 20	78

Math Help for College

Summary 2

Summary 2 Pretest 2
(25 minutes)

Name

Read directions carefully. Answers must be in the indicated place.

Show all steps for full credits. **Each problem is worth 5 points.**

Do only 1 operation per 1 step. **Solve 1 problem 5 minutes**

1) **Multiply** $\begin{array}{r} 706 \\ \underline{\times 605} \end{array}$

1)_______

2) **Divide** $243714 \div 302$

2)______

3) **Evaluate** $240240 \div 80 \times 3$

3)______

4) **Evaluate** $560560 - 480480 \div 8 \times 5$

4)______

5) **Evaluate** $120 \div 3 \times 2 - 60 \div 4 \times 3$

5)______

Summary 2

Chapter 1 Addition or Subtraction Numbers (20-29)

Summary 2 Chapter 1 Contents
Addition or Subtraction Numbers (20–29)
Lesson 1 - 11
Basic Problems

Work with lesson 1, read rules and examples, 2) Solve the problems from contents, 3) Check answers in the end of chapter 1, 4) Work with the next lessons 2, 3, 4, 5, …, 5) Solve Test 1 A, 6) Grading Scale: A =10 points, B=9points, C=8 points, D=7 points, F=0-6 points, 7) If you solve 10 problems correctly, you get an A, 8) If have an A, you can move on to next chapter, 9) If your grade is lower than an A, work with the mistakes, then solve Test 1B, 10) When you finish all of the chapters, solve Final Exam in last chapter.

Summary 2 Chapter 1 Lesson 1

Numbers from 0 to 9

Count forward from 1 to 9 — 1, 2, 3, 4, 5, 6, 7, 8, 9

Count backwards from 9 to 1 — 9, 8, 7, 6, 5, 4, 3, 2, 1

The smallest number is 0 — The biggest number is 9

Equal means 5 = 5 — Number zero is less than one 0 < 1

Less than means (<)

0 < 1	1 < 2	2 < 3	3 < 4	4 < 5
5 < 6	6 < 7	7 < 8	8 < 9	9 = 9

More than means (>)

9 > 8	8 > 7	7 > 6	6 > 5	5 > 4
4 > 3	3 > 2	2 > 1	1 > 0	0 = 0

Review 1 A

Add or Subtract numbers in row

	a	b	c	d	e
# 1	2 + 0	3 + 3	8 + 0	0 + 6	7 + 1
# 2	1 + 8	2 + 2	7 + 2	2 + 4	6 + 2
# 3	9 - 9	8 – 7	7 – 6	7 – 5	8 – 4
# 4	4 – 4	6 – 3	9 – 2	5 – 2	9 – 1

Add or Subtract numbers in column

	a	b	c	d	e
# 5	1 + 4	6 + 2	3 + 3	3 + 4	5 + 3
# 6	3 + 6	2 + 4	4 + 3	4 + 4	4 + 5
# 7	9 -8	7 - 7	6 - 6	6 - 5	7 - 4
# 8	9 -3	5 - 3	8 - 2	4 - 2	8 - 1

Summary 2 Chapter 1 Lesson 1

Addition numbers in row

The first operation with number is addition +

Property of addition

Two numbers can add in any order.
The sum wills the same. 2 + 3 = 5 3 + 2 = 5
The sum of a number and zero is the same as the number
5 + 0 = 5 0 + 5 = 5

Subtracting numbers in row

Second operation with numbers is subtraction –
Any number subtract zero result is the same number 5 – 0 = 5
Any number subtract the same number result is zero 5 – 5 = 0

Review 1 B

Add or Subtract numbers in row

	a	b	c	d	e
# 1	1 + 6	4 + 5	1 + 5	5 + 2	3 + 5
# 2	5 + 4	0 + 6	6 + 1	2 + 6	6 + 3
# 3	8 - 8	9 – 6	9 – 5	5 – 5	6 – 4
# 4	8 – 3	4 – 3	7 – 2	3 – 2	7 – 0

Add or Subtract numbers in column

	a	b	c	d	e
# 5	6 + 2	5 + 3	1 + 7	7 + 2	1 + 8
# 6	7 + 1	9 + 0	2 + 7	8 + 1	0 + 9
# 7	9 -7	8 - 6	8 - 5	9 - 4	5 - 4
# 8	7 -3	3 - 3	6 - 2	2 - 2	6 - 0

Summary 2 Chapter 1 Lesson 2

Two-digit numbers

One-digit numbers

One-digit numbers have one digit. This is numbers from 0 to 9
0, 1, 2, 3, 4, 5, 6, 7, 8, 9 Digits 0, 1, 2, 3, 4, 5, 6, 7, 8, 9

Two-digit numbers

Two-digit numbers have two digits. This is numbers from 10 to 99
Digit on the left are called tens. Digit on the right are called ones

Comparison numbers

One-digit numbers always less than two-digit numbers 9 < 11
Two-digit numbers always greater than one-digit numbers 19 > 8
Compare the tens and then compare the ones 12 < 18 18 > 12

Review 1 A

Add or Subtract numbers in row

	a	b	c	d	e
# 1	8 + 5	6 + 8	8 + 7	8 + 8	9 + 8
# 2	9 + 5	6 + 9	9 + 7	8 + 9	9 + 9
# 3	13 - 4	13 – 6	13 – 8	14 – 5	14 – 7
# 4	15 – 6	15 – 8	16 – 7	17 – 8	16 – 9

Add or Subtract numbers in column

	a	b	c	d	e
# 5	9 + 9	8 + 9	9 + 8	7 + 9	8 + 8
# 6	9 + 7	6 + 9	8 + 7	7 + 8	9 + 6
# 7	18 -9	17 - 9	16 - 8	15 - 6	15 - 8
# 8	14 -5	14 - 7	14 - 9	13 - 6	13 - 8

Summary 2 Chapter 1 Lesson 2

Rewrite and Memorize two-digit numbers

Numbers from 10 to 19	10, 11, 12, 13, 14, 15, 16, 17, 18, 19
Numbers from 20 to 29	20, 21, 22, 23, 24, 25, 26, 27, 28, 29
Numbers from 30 to 39	30, 31, 32, 33, 34, 35, 36, 37, 38, 39
Numbers from 40 to 49	40, 41, 42, 43, 44, 45, 46, 47, 48, 49
Numbers from 50 to 59	50, 51, 52, 53, 54, 55, 56, 57, 58, 59
Numbers from 60 to 69	60, 61, 62, 63, 64, 65, 66, 67, 68, 69
Numbers from 70 to 79	70, 71, 72, 73, 74, 75, 76, 77, 78, 79
Numbers from 80 to 89	80, 81, 82, 83, 84, 85, 86, 87, 88, 89
Numbers from 90 to 99	90, 91, 92, 93, 94, 95, 96, 97, 98, 99

Review 1 B

Add or Subtract numbers in row

	a	b	c	d	e
# 1	5 + 8	8 + 6	7 + 8	8 + 8	8 + 9
# 2	5 + 9	9 + 6	7 + 9	9 + 8	9 + 9
# 3	13 - 5	13 – 7	13 – 9	14 – 6	14 – 8
# 4	15 – 7	15 – 9	16 – 8	17 – 9	18 – 9

Add or Subtract numbers in column

	a	b	c	d	e
# 5	9 + 5	8 + 6	7 + 7	6 + 8	5 + 9
# 6	8 + 5	7 + 6	6 + 7	5 + 8	4 + 9
# 7	17 -8	16 - 7	16 - 9	15 - 7	15 - 9
# 8	14 -6	14 - 8	13 - 5	13 - 7	13 - 9

Summary 2 Chapter 1 Lesson 3

Three Numbers

Addition three numbers in row

Three numbers can add in any order

9 + 4 + 3 = 13 + 3 = 16 9 + 4 + 3 = 9 + 7 = 16

Subtraction three numbers in row

To subtract numbers, work from left to right

10 – 1 – 4 = 9 – 4 = 5 10 – 1 – 5 = 9 – 5 = 4

Two operations with three numbers

To do two operations, work from left to right

5 + 6 – 9 = 11 – 9 = 2 9 – 8 + 7 = 1 + 7 = 8

5 + 8 – 9 = 13 – 9 = 4 9 – 8 + 5 = 1 + 5 = 6

Review 1 A

Add and Subtract numbers in row

	a	b	c	d	e
# 1	9 + 6 + 4	8 + 6 + 5	10 – 3 - 1	10 – 4 - 2	6 + 6 - 2
# 2	6 + 7 - 3	10 - 4 + 3	10 – 5 + 4	10 + 10	20 + 20

Add and Subtract numbers in column

	a	b	c	d	e
# 3	10 + 20	20 - 10	11 + 18	55 - 41	15 + 13 11
# 4	20 + 30	80 - 50	12 + 16	44 - 31	14 + 11 12

Summary 2 Chapter 1 Lesson 3

Adding or Subtraction numbers in column

Add or Subtract the ones and then add or subtract the tens

80	90	27	96	23
+ 10	- 10	+ 52	- 25	+ 32
90	80	79	71	21
				76

Review 1 B

Add and Subtract Numbers in row

	a	b	c	d	e
# 1	7 + 5 + 4	6 + 4 + 3	10 – 5 - 3	10 – 6 - 4	6 + 8 - 4
# 2	6 + 9 - 5	10 – 6 + 5	10 – 7 + 6	40 + 30	60 + 20

Add and Subtract Numbers in column

	a	b	c	d	e
# 3	30	70	73	74	42
	+ 40	- 30	+ 24	- 21	+ 20
					34
# 4	60	60	84	62	51
	+ 20	- 10	+12	- 11	+ 13
					23

Summary 2 Chapter 1 Lesson 4

Add or Subtract numbers in column

5 tens – 1 ten = 4 tens   10 ones = 1 ten

1) Number 50

1	4 10	1	4 10
a) 29	b) 5 0	c) 19	d) 50
+21	- 2 1	+31	- 19
50	2 9	50	31

2) Number 51

1	4 11	1	4 11
a) 29	b) 5 1	c) 19	d) 51
+22	- 2 2	+32	- 19
51	2 9	51	32

3) Number 52

1	4 12	1	4 12
a) 29	b) 5 2	c) 19	d) 52
+23	- 2 3	+33	- 19
52	2 9	52	33

4) Number 53

1	4 13	1	4 13
a) 29	b) 5 3	c) 19	d) 53
+24	- 2 4	+34	- 19
53	2 9	53	34

5) Numbers 54

1	4 14	1	4 14
a) 29	b) 5 4	c) 19	d) 54
+25	- 2 5	+ 35	- 19
54	2 9	54	35

Summary 2 Chapter 1 Lesson 4

Add or Subtract numbers in column

5 tens – 1 ten = 4 tens   10 ones = 1 ten

6) Number 55

1	4 15	1	4 15
a) 29	b) 5 5	c) 19	d) 55
+26	- 2 6	+36	- 19
55	2 9	55	36

7) Number 56

1	4 16	1	4 16
a) 29	b) 5 6	c) 19	d) 56
+27	- 2 7	+37	- 19
56	2 9	56	37

8) Number 57

1	4 17	1	4 17
a) 29	b) 5 7	c) 19	d) 57
+28	- 2 8	+38	- 19
57	2 9	57	38

9) Number 58

1	4 18	1	4 18
a) 29	b) 5 8	c) 19	d) 58
+29	- 2 9	+39	- 19
58	2 9	58	39

10) Number 59

a) 29	b) 59	c) 19	d) 59
+ 30	- 30	+40	-19
59	29	59	40

Summary 2 Chapter 1 Lesson 5

Addition and Subtraction Numbers

Review 1 A

Add or Subtract numbers in row

	a	b	c	d	e
# 1	8 + 5	6 + 8	8 + 7	8 + 8	9 + 8
# 2	9 + 5	6 + 9	9 + 7	8 + 9	9 + 9
# 3	13 - 4	13 – 6	13 – 8	14 – 5	14 – 7
# 4	15 – 6	15 – 8	16 – 7	17 – 8	16 – 9

Add or Subtract numbers in column

	a	b	c	d	e
# 5	9 + 9	8 + 9	9 + 8	7 + 9	8 + 8
# 6	9 + 7	6 + 9	8 + 7	7 + 8	9 + 6
# 7	18 -9	17 - 9	16 - 8	15 - 6	15 - 8
# 8	14 -5	14 - 7	14 - 9	13 - 6	13 - 8

Review 1 B

Add and Subtract numbers in row

	a	b	c	d	e
# 1	8 + 6 + 4	9 + 4 + 5	10 – 4 - 1	10 – 5 - 1	7 + 6 - 2
# 2	8 + 7 - 3	12 - 4 + 3	11 – 5 + 4	10 + 20	20 + 20

Add and Subtract numbers in column

	a	b	c	d	e
# 3	10 + 20	80 - 70	12 + 16	55 - 31	14 + 12 11
# 4	20 + 30	60 - 40	13 + 16	54 - 31	15 + 12 12

Summary 2 Chapter 1 Lesson 6

Number 20

Remember

1+ 9 = 2 + 8 = 3 + 7 = 4 + 6 = 5 + 5 = 10

Add or Subtract numbers in row

6 + 14 = 14 + 6 =10+4+6=10+10= 20

20 – 14 = 20 – 10 – 4 = 10 – 4 = 6

Review 1 A

	a	b	c	d
# 1	20 + 0	20 - 0	10 + 10	20 - 10
# 2	19 + 1	20 – 1	9 + 11	20 – 11
# 3	18 + 2	20 - 2	8 + 12	20 - 12
# 4	17 + 3	20 - 3	7 + 13	20 - 13
# 5	16 + 4	20 – 4	6 + 14	20 – 14
# 6	15 + 5	20 - 5	5 + 15	20 - 15
# 7	14 + 6	20 – 6	4 + 16	20 – 16
# 8	13 + 7	20 - 7	3 + 17	20 - 17
# 9	12 + 8	20 – 8	2 + 18	20 – 18
# 10	11 + 9	20 – 9	1 + 19	20 - 19

Number 21

Add or Subtract numbers in row

7 + 14 = 14 + 7 = 10 + 4 + 7 =10 + 11 = 21

21 – 14 = 21 – 11 – 3 = 10 – 3 = 7

Review 1 B

	a	b	c	d
# 11	21 + 0	21 - 0	11 + 10	21 - 10
# 12	20 + 1	21 – 1	10 + 11	21 – 11
# 13	19 + 2	21 - 2	9 + 12	21 - 12
# 14	18 + 3	21 – 3	8 + 13	21 – 13
# 15	17 + 4	21 - 4	7 + 14	21 – 14
# 16	16 + 5	21 – 5	6 + 15	21 – 15
# 17	15 + 6	21 - 6	5 + 16	21 - 16
# 18	14 + 7	21 – 7	4 + 17	21 – 17
# 19	13 + 8	21 - 8	3 + 18	21 - 18
# 20	12 + 9	21 – 9	2 + 19	21 – 19

Summary 2 Chapter 1 Lesson 7

Number 22

Remember

1+ 9 = 2 + 8 = 3 + 7 = 4 + 6 = 5 + 5 = 10

Add or Subtract numbers in row

4 + 18 = 18 + 4 = 10 + 8 + 4 = 10 + 12 = 22

22 – 14 = 22 – 12 – 2 = 10 – 2 = 8

Review 1 A

	a	b	c	d
# 1	22 + 0	22 - 0	12 + 10	22 - 10
# 2	21 + 1	22 –1	11 + 11	22 – 11
# 3	20 + 2	22 - 2	10 + 12	22 - 12
# 4	19 + 3	22 –3	9 + 13	22 – 13
# 5	18 + 4	22 - 4	8 + 14	22 – 14
# 6	17 + 5	22 –5	7 + 15	22 – 15
# 7	16 + 6	22 - 6	6 + 16	22 - 16
# 8	15 + 7	22 –7	5 + 17	22 – 17
# 9	14 + 8	22 - 8	4 + 18	22 - 18
# 10	13 + 9	22 –9	3 + 19	22 – 19

Number 23

9 + 14 = 14 + 9 = 10 + 4 + 9=10+13 = 23

23 – 14 = 23 – 13 – 1 = 10 – 1 = 9

Review 1 B

	a	b	c	d
# 11	23 + 0	23 - 0	13 + 10	23 - 10
# 12	22 + 1	23 – 1	12 + 11	23 – 11
# 13	21 + 2	23 - 2	11 + 12	23 - 12
# 14	20 + 3	23 – 3	10 + 13	23 – 13
# 15	19 + 4	23 - 4	9 + 14	23 – 14
# 16	18 + 5	23 – 5	8 + 15	23 –15
# 17	17 + 6	23 - 6	7 + 16	23 - 16
# 18	16 + 7	23 – 7	6 + 17	23 –17
# 19	15 + 8	23 - 8	5 + 18	23 - 18
# 20	14 + 9	23 – 9	4 + 19	23 -19

Summary 2 Chapter 1 Lesson 8

Remember

1+ 9 = 2 + 8 = 3 + 7 = 4 + 6 = 5 + 5 = 10

Number 24

10 + 14 = 10 + 10 + 4 = 20 + 4 = 24

24 – 14 = 24 – 10 – 4 = 14 – 4 = 10

Review 1 A

Add or Subtract numbers in row

	a	b	c	d
# 1	24 + 0	24 - 0	14 + 10	24 - 10
# 2	23 + 1	24 – 1	13 + 11	24 – 11
# 3	22 + 2	24 - 2	12 + 12	24 - 12
# 4	21 + 3	24 – 3	11 + 13	24 – 13
# 5	20 + 4	24 - 4	10 + 14	24 – 14
# 6	19 + 5	24 – 5	9 + 15	24 – 15
# 7	18 + 6	24 - 6	8 + 16	24 - 16
# 8	17 + 7	24 – 7	7 + 17	24 – 17
# 9	16 + 8	24 - 8	6 + 18	24 - 18
# 10	15 + 9	24 – 9	5 + 19	24 – 19

Number 25

11 + 14 = 10 + 1 + 10 + 4 = 20 + 5 = 25

25 – 14 = 25 – 10 – 4 = 15 – 4 = 11

Review 1 B

Add or Subtract numbers in row

	a	b	c	d
# 11	25 + 0	25 - 0	15 + 10	25 - 10
# 12	24 + 1	25 – 1	14 + 11	25 –11
# 13	23 + 2	25 - 2	13 + 12	25 - 12
# 14	22 + 3	25 – 3	12 + 13	25 –13
# 15	21 + 4	25 - 4	11 + 14	25 – 14
# 16	20 + 5	25 – 5	10 + 15	25 –15
# 17	19 + 6	25 - 6	9 + 16	25 - 16
# 18	18 + 7	25 – 7	8 + 17	25 –17
# 19	17 + 8	25 - 8	7 + 18	25 - 18
# 20	16 + 9	25 – 9	6 + 19	25 –19

Summary 2 Chapter 1 Lesson 9

Remember

1+ 9 = 2 + 8 = 3 + 7 = 4 + 6 = 5 + 5 = 10

Numbers 26

12 + 14 = 10 + 2 + 10 + 4 = 20 + 6 = 26

26 - 14 = 26 – 10 – 4 = 16 – 4 = 12

Review 1 A

Add or Subtract numbers in row

	a	b	c	d
# 1	26 + 0	26 - 0	16 + 10	26 - 10
# 2	25 + 1	26 – 1	15 + 11	26 – 11
# 3	24 + 2	26 - 2	14 + 12	26 - 12
# 4	23 + 3	26 – 3	13 + 13	26 – 13
# 5	22 + 4	26 - 4	12 + 14	26 – 14
# 6	21 + 5	26 – 5	11 + 15	26 – 15
# 7	20 + 6	26 - 6	10 + 16	26 - 16
# 8	19 + 7	26 – 7	9 + 17	26 – 17
# 9	18 + 8	26 - 8	8 + 18	26 - 18
# 10	17 +9	26 – 9	7 + 19	26 – 19

Number 27

13 + 14 = 10 + 3 + 10 + 4 = 20 + 7 = 27

27 – 14 = 27 – 10 – 4 = 17 – 4 = 13

Review 1 B

Add or Subtract numbers in row

	a	b	c	d
# 11	27 + 0	27 - 0	17 + 10	27 - 10
# 12	26 + 1	27 – 1	16 + 11	27 –11
# 13	25 + 2	27 - 2	15 + 12	27 - 12
# 14	24 + 3	27 – 3	14 + 13	27 –13
# 15	23 + 4	27 - 4	13 + 14	27 – 14
# 16	22 + 5	27 – 5	12 + 15	27 –15
# 17	21 + 6	27 - 6	11 + 16	27 - 16
# 18	20 + 7	27 – 7	10 + 17	27 –17
# 19	19 + 8	27 - 8	9 + 18	27 - 18
# 20	18 + 9	27 – 9	8 + 19	27 - 19

Summary 2 Chapter 1 Lesson 10

Remember

1+ 9 = 2 + 8 = 3 + 7 = 4 + 6 = 5 + 5 = 10

Number 28

14 + 14 = 10 + 4 + 10 + 4 = 20 + 8 = 28

28 – 14 = 28 – 10 – 4 = 18 – 4 = 14

Review 1 A

Add or Subtract numbers in row

	a	b	c	d
# 1	28 + 0	28 - 0	18 + 10	28 - 10
# 2	27 + 1	28 – 1	17 + 11	28 – 11
# 3	26 + 2	28 - 2	16 + 12	28 - 12
# 4	25 + 3	28 – 3	15 + 13	28 – 13
# 5	24 + 4	28 - 4	14 + 14	28 – 14
# 6	23 + 5	28 – 5	13 + 15	28 –15
# 7	22 + 6	28 - 6	12 + 16	28 – 16
# 8	21 + 7	28 – 7	11 + 17	28 –17
# 9	20 + 8	28 - 8	10 + 18	28 – 18
# 10	19 + 9	28 – 9	9 + 19	28 –19

Number 29

15 + 14 = 10 + 5 + 10 + 4 = 20 + 9 = 29

29 – 14 = 29 – 10 – 4 = 19 – 4 = 15

Review 1 B

Add or Subtract numbers in row

	a	b	c	d
# 11	29 + 0	29 - 0	19 + 10	29 - 10
# 12	28 + 1	29 – 1	18 + 11	29 – 11
# 13	27 + 2	29 - 2	17 + 12	29 - 12
# 14	26 + 3	29 – 3	16 + 13	29 – 13
# 15	25 + 4	29 - 4	15 + 14	29 - 14
# 16	24 + 5	29 – 5	14 + 15	29 – 15
# 17	23 + 6	29 - 6	13 + 16	29 - 16
# 18	22 + 7	29 – 7	12 + 17	29 – 17
# 19	21 + 8	29 - 8	11 + 18	29 - 18
# 20	20 + 9	29 – 9	10 + 19	29 – 19

Summary 2 Chapter 1 Lesson 11

Addition or Subtraction Numbers in row

Numbers 20 to 29

Test 1 A (50 minutes)

Add or Subtract Numbers in row

	a	b	c	d
# 1	20 + 0	20 - 0	10 + 10	20 - 10
# 2	19 + 1	20 – 1	9 + 11	20 – 11
# 3	19 + 2	21 - 2	9 + 12	21 - 12
# 4	18 + 3	21 – 3	8 + 13	21 – 13
# 5	18 + 4	22 - 4	8 + 14	22 - 14
# 6	17 + 5	22 –5	7 + 15	22 – 15
# 7	17 + 6	23 - 6	7 + 16	23 - 16
# 8	16 + 7	23 – 7	6 + 17	23 –17
# 9	16 + 8	24 - 8	6 + 18	24 - 18
# 10	15 + 9	24 – 9	5 + 19	24 – 19

Summary 2 Chapter 1 Lesson 11

Addition or Subtraction Numbers in row

Numbers 20 to 29

Test 1 B (50 minutes)

Add or Subtract Numbers in row

	a	b	c	d
# 1	17 + 8	25 - 8	7 + 18	25 - 18
# 2	16 + 9	25 – 9	6 + 19	25 –19
# 3	26 + 0	26 - 0	16 + 10	26 - 10
# 4	25 + 1	26 – 1	15 + 11	26 – 11
# 5	25 + 2	27 - 2	15 + 12	27 - 12
# 6	24 + 3	27 – 3	14 + 13	27 –13
# 7	26 + 2	28 - 2	16 + 12	28 - 12
# 8	25 + 3	28 – 3	15 + 13	28 – 13
# 9	25 + 4	29 - 4	15 + 14	29 - 14
# 10	24 + 5	29 – 5	14 + 15	29 – 15

Summary 2 Answers Chapter 1

Pretest 2 1) 427,130 2) 807 3) 9009 4) 260, 260 5) 35

Lesson 1

Review 1 A

	a	b	c	d	e		a	b	c	d	e
# 1	2	6	8	6	8	# 5	5	8	6	7	8
# 2	9	4	9	6	8	# 6	9	6	7	8	9
# 3	0	1	1	2	4	# 7	1	0	0	1	3
# 4	0	3	7	3	8	# 8	6	2	6	2	7

Review 1 B

	a	b	c	d	e		a	b	c	d	e
# 1	7	9	6	7	8	# 5	8	8	8	9	9
# 2	9	6	7	8	9	# 6	8	9	9	9	9
# 3	0	3	4	0	2	# 7	2	2	3	5	1
# 4	5	1	5	1	7	# 8	4	0	4	0	6

Lesson 2

Review 1 A

	a	b	c	d	e		a	b	c	d	e
# 1	13	14	15	16	17	# 5	18	17	17	16	16
# 2	14	15	16	17	18	# 6	16	15	15	15	15
# 3	9	7	5	9	7	# 7	9	8	8	9	7
# 4	9	7	9	9	7	# 8	9	7	5	7	5

Review 1 B

	a	b	c	d	e		a	b	c	d	e
# 1	13	14	15	16	17	# 5	14	14	14	14	14
# 2	14	15	16	17	18	# 6	13	13	13	13	13
# 3	8	6	4	8	6	# 7	9	9	7	8	6
# 4	8	6	8	8	9	# 8	8	6	8	6	4

Lesson 3

Review 1 A

	a	b	c	d	e		a	b	c	d	e
# 1	19	19	6	4	10	# 3	30	10	29	14	39
# 2	10	9	9	20	40	# 4	50	30	28	13	37

Review 1 B

	a	b	c	d	e		a	b	c	d	e
# 1	16	13	2	0	10	# 3	70	40	97	53	96
# 2	10	9	9	70	80	# 4	80	50	96	51	87

Summary 2 Answers Chapter 1

Lesson 5

Review 1 A

	a	b	c	d	e		a	b	c	d	e
# 1	13	14	15	16	17	# 5	18	17	17	16	16
# 2	14	15	16	17	18	# 6	16	15	15	15	15
# 3	9	7	5	9	7	# 7	9	8	8	9	7
# 4	9	7	9	9	7	# 8	9	7	5	7	5

Review 1 B

	a	b	c	d	e		a	b	c	d	e
# 1	18	18	5	4	11	# 3	30	10	28	24	37
# 2	12	11	10	30	40	# 4	50	20	29	23	39

Lesson 6 Review 1 A / Review 1 B

	a	b	c	d		a	b	c	d
# 1	20	20	20	10	# 11	21	21	21	11
# 2	20	19	20	9	# 12	21	20	21	10
# 3	20	18	20	8	# 13	21	19	21	9
# 4	20	17	20	7	# 14	21	18	21	8
# 5	20	16	20	6	# 15	21	17	21	7
# 6	20	15	20	5	# 16	21	16	21	6
# 7	20	14	20	4	# 17	21	15	21	5
# 8	20	13	20	3	# 18	21	14	21	4
# 9	20	12	20	2	# 19	21	13	21	3
# 10	20	11	20	1	# 20	21	12	21	2

Summary 2 Answers Chapter 1

Lesson 7 Review 1 A / **Review 1 B**

	a	b	c	d		a	b	c	d
# 1	22	22	22	12	# 11	23	23	23	13
# 2	22	21	22	11	# 12	23	22	23	12
# 3	22	20	22	10	# 13	23	21	23	11
# 4	22	19	22	9	# 14	23	20	23	10
# 5	22	18	22	8	# 15	23	19	23	9
# 6	22	17	22	7	# 16	23	18	23	8
# 7	22	16	22	6	# 17	23	17	23	7
# 8	22	15	22	5	# 18	23	16	23	6
# 9	22	14	22	4	# 19	23	15	23	5
# 10	22	13	22	3	# 20	23	14	23	4

Lesson 8 Review 1 A / **Review 1 B**

	a	b	c	d		a	b	c	d
# 1	24	24	24	14	# 11	25	25	25	15
# 2	24	23	24	13	# 12	25	24	25	14
# 3	24	22	24	12	# 13	25	23	25	13
# 4	24	21	24	11	# 14	25	22	25	12
# 5	24	20	24	10	# 15	25	21	25	11
# 6	24	19	24	9	# 16	25	20	25	10
# 7	24	18	24	8	# 17	25	19	25	9
# 8	24	17	24	7	# 18	25	18	25	8
# 9	24	16	24	6	# 19	25	17	25	7
# 10	24	15	24	5	# 20	25	16	25	6

Lesson 9 Review 1 A / **Review 1 B**

	a	b	c	d		a	b	c	d
# 1	26	26	26	16	# 11	27	27	27	17
# 2	26	25	26	15	# 12	27	26	27	16
# 3	26	24	26	14	# 13	27	25	27	15
# 4	26	23	26	13	# 14	27	24	27	14
# 5	26	22	26	12	# 15	27	23	27	13
# 6	26	21	26	11	# 16	27	22	27	12
# 7	26	20	26	10	# 17	27	21	27	11
# 8	26	19	26	9	# 18	27	20	27	10
# 9	26	18	26	8	# 19	27	19	27	9
# 10	26	17	26	7	# 20	27	18	27	8

Summary 2 Answers Chapter 1

Lesson 10

Review 1 A | **Review 1 B**

	a	b	c	d		a	b	c	d
# 1	28	28	28	18	# 11	29	29	29	19
# 2	28	27	28	17	# 12	29	28	29	18
# 3	28	26	28	16	# 13	29	27	29	17
# 4	28	25	28	15	# 14	29	26	29	16
# 5	28	24	28	14	# 15	29	25	29	15
# 6	28	23	28	13	# 16	29	24	29	14
# 7	28	22	28	12	# 17	29	23	29	13
# 8	28	21	28	11	# 18	29	22	29	12
# 9	28	20	28	10	# 19	29	21	29	11
# 10	28	19	28	9	# 20	29	20	29	10

Lesson 11

Test 1 A

	a	b	c	d		a	b	c	d
# 1	20	20	20	10	# 6	22	17	22	7
# 2	20	19	20	9	# 7	23	17	23	7
# 3	21	19	21	9	# 8	23	16	23	6
# 4	21	18	21	8	# 9	24	16	24	6
# 5	22	18	22	8	# 10	24	15	24	5

Test 1 B

	a	b	c	d		a	b	c	d
# 1	25	17	25	7	# 6	27	24	27	14
# 2	25	16	25	6	# 7	28	26	28	16
# 3	26	26	26	16	# 8	28	25	28	15
# 4	26	25	26	15	# 9	29	25	29	15
# 5	27	25	27	15	# 10	29	24	29	14

Summary 2

Chapter 2 Addition or Subtraction Numbers (50-59)

Summary 2 Chapter 2 Contents
Addition or Subtraction Numbers (50 – 59)
Lesson 1 - 11
Basic Problems

Work with lesson 1, read rules and examples, 2) Solve the problems from contents, 3) Check answers in the end of chapter 1, 4) Work with the next lessons 2, 3, 4, 5, ..., 5) Solve Test 1 A, 6) Grading Scale: A =10 points, B=9points, C=8 points, D=7 points, F=0-6 points, 7) If you solve 10 problems correctly, you get an A, 8) If have an A, you can move on to next chapter, 9) If your grade is lower than an A, work with the mistakes, then solve Test 1B, 10) When you finish all of the chapters, solve Final Exam in last chapter.

Summary 2 Chapter 2 Lesson 1
Number 50
Add or Subtract numbers in column
5 tens – 1 ten = 4 tens 10 ones = 1 ten

a)	b)	c)	d)
1	4 10	1	4 10
29	5 0	19	50
+21	- 2 1	+31	- 19
50	2 9	50	31

Add or Subtract

	a	b	c	d	e
# 1	50 + 0	50 - 0	30 + 20	50 - 20	10 + 40
# 2	49 + 1	50 - 1	29 +21	50 - 21	9 + 41
# 3	48 + 2	50 - 2	28 + 22	50 - 22	8 + 42
# 4	47 + 3	50 - 3	27 + 23	50 - 23	7 + 43
# 5	46 + 4	50 - 4	26 + 24	50 - 24	6 + 44
# 6	45 + 5	50 - 5	25 + 25	50 - 25	5 + 45
# 7	44 + 6	50 - 6	24 + 26	50 - 26	4 + 46
# 8	43 + 7	50 - 7	23 + 27	50 - 27	3 + 47
# 9	42 + 8	50 - 8	22 + 28	50 - 28	2 + 48
# 10	41 + 9	50 - 9	21 + 29	50 - 29	1 + 49

Summary 2 Chapter 2 Lesson 2
Number 51
Add or Subtract numbers in column
5 tens – 1 ten = 4 tens   1 ten = 10 ones

a)	b)	c)	d)
1	4 11	1	4 11
29	5 1	19	51
+22	- 2 2	+32	- 19
51	2 9	51	32

Add or Subtract

	a	b	c	d	e
# 1	51 +0	51 - 0	31 +20	51 -20	11 +40
# 2	50 +1	51 -1	30 +21	51 -21	10 + 41
# 3	49 + 2	51 - 2	29 +22	51 -22	9 +42
# 4	48 + 3	51 - 3	28 +23	51 - 23	8 +43
# 5	47 + 4	51 -4	27 +24	51 -24	7 +44
# 6	46 + 5	51 -5	26 +25	51 -25	6 +45
# 7	45 + 6	51 - 6	25 + 26	51 - 26	5 + 46
# 8	44 + 7	51 - 7	24 +27	51 -27	4 + 47
# 9	43 + 8	51 - 8	23 + 28	51 -28	3 +48
# 10	42 + 9	51 - 9	22 +29	51 -29	2 +49

Summary 2 Chapter 2 Lesson 3
Number 52

5 tens – 1 ten = 4 tens   1 ten = 10 ones

a)	b)	c)	d)
1	4 12	1	4 12
29	5 2	19	52
+23	- 2 3	+33	- 19
52	2 9	52	33

Add or Subtract numbers in column

	a	b	c	d	e
# 1	52 +0	52 -0	32 +20	52 -20	12 +40
# 2	51 +1	52 -1	31 +21	52 -21	11 +41
# 3	50 +2	52 -2	30 +22	52 - 22	10 +42
# 4	49 +3	52 -3	29 +23	52 -23	9 +43
# 5	48 +4	52 -4	28 +24	52 -24	8 +44
# 6	47 +5	52 -5	27 +25	52 -25	7 +45
# 7	46 +6	52 -6	26 +26	52 -26	6 +46
# 8	45 +7	52 - 7	25 +27	52 -27	5 +47
# 9	44 +8	52 - 8	24 +28	52 -28	4 +48
# 10	43 +9	52 - 9	23 +29	52 -29	3 +49

Summary 2 Chapter 2 Lesson 4
Number 53

5 tens – 1 ten = 4 tens   1 ten = 10 ones

a)	b)	c)	d)
1	4 13	1	4 13
29	5 3	19	53
+24	- 2 4	+34	- 19
53	2 9	53	34

Add or Subtract numbers in column

	a	b	c	d	e
# 1	53 + 0	53 - 0	33 +20	53 -20	13 +40
# 2	52 + 1	53 -1	32 +21	53 -21	12 +41
# 3	51 + 2	53 - 2	31 +22	53 -22	11 +42
# 4	50 + 3	53 - 3	30 +23	53 -23	10 +43
# 5	49 + 4	53 - 4	29 +24	53 -24	9 +44
# 6	48 + 5	53 - 5	28 +25	53 -25	8 +45
# 7	47 + 6	53 - 6	27 +26	53 -26	7 +46
# 8	46 + 7	53 - 7	26 +27	53 -27	6 +47
# 9	45 + 8	53 - 8	25 +28	53 -28	5 +48
# 10	44 +9	53 - 9	24 +29	53 -29	4 +49

Summary 2 Chapter 2 Lesson 5
Number 54
Add or Subtract numbers in column
5 tens – 1 ten = 4 tens   1 ten = 10 ones

a)	b)	c)	d)
1	4 14	1	4 14
29	5 4	19	54
+25	- 2 5	+35	- 19
54	2 9	54	35

Add or Subtract numbers in column

	a	b	c	d	e
# 1	54 +0	54 -0	34 +20	54 -20	14 +40
# 2	53 +1	54 -1	33 +21	54 -21	13 +41
# 3	52 +2	54 -2	32 +22	54 -22	12 +42
# 4	51 +3	54 -3	31 +23	54 -23	11 +43
# 5	50 +4	54 -4	30 +24	54 -24	10 +44
# 6	49 +5	54 -5	29 +25	54 -25	9 +45
# 7	48 +6	54 -6	28 +26	54 -26	8 +46
# 8	47 +7	54 -7	27 +27	54 -27	7 +47
# 9	46 +8	54 -8	26 +28	54 -28	6 +48
# 10	45 +9	54 -9	25 +29	54 -29	7 +47

Summary 2 Chapter 2 Lesson 6
Number 55
5 tens – 1 ten = 4 tens   1 ten = 10 ones

a)	b)	c)	d)
1	4 15	1	4 15
29	5 5	19	55
+26	- 2 6	+36	- 19
55	2 9	55	36

Add or Subtract numbers in column

	a	b	c	d	e
# 1	55 +0	55 -0	35 +20	55 -20	15 +40
# 2	54 +1	55 -1	34 +21	55 -21	14 +41
# 3	53 +2	55 -2	33 +22	55 -22	13 +42
# 4	52 +3	55 -3	32 +23	55 -23	12 +43
# 5	51 +4	55 -4	31 +24	55 -24	11 +44
# 6	50 +5	55 -5	30 + 25	55 -25	10 +45
# 7	49 +6	55 -6	29 +26	55 -26	9 +46
# 8	48 +7	55 -7	28 + 27	55 - 27	8 +47
# 9	47 +8	55 -8	27 +28	55 -28	7 +48
# 10	46 +9	55 -9	26 +29	55 -29	6 +49

Summary 2 Chapter 2 Lesson 7
Number 56
Add or Subtract numbers in column
5 tens – 1 ten = 4 tens   1 ten = 10 ones

a)	b)	c)	d)
1	4 16	1	4 16
29	5 6	19	56
+27	- 2 7	+37	- 19
56	2 9	56	37

Add or Subtract numbers in column

	a	b	c	d	e
# 1	56 +0	56 -0	36 +20	56 -20	16 +40
# 2	55 +1	56 -1	35 +21	56 -21	15 +41
# 3	54 +2	56 -2	34 +22	56 -22	14 +42
# 4	53 +3	56 -3	33 +23	56 -23	13 +43
# 5	52 +4	56 -4	32 +24	56 -24	12 +44
# 6	51 +5	56 -5	31 + 25	56 -25	11 +45
# 7	50 +6	56 -6	30 +26	56 -26	10 +46
# 8	49 +7	56 -7	29 +27	56 -27	9 +47
# 9	48 +8	56 -8	28 +28	56 -28	8 +48
# 10	47 +9	56 -9	27 +29	56 -29	7 + 49

Summary 2 Chapter 2 Lesson 8
Number 57
5 tens – 1 ten = 4 tens   1 ten = 10 ones

a)	b)	c)	d)
1	4 17	1	4 17
29	5 7	19	57
+28	- 2 8	+38	- 19
57	2 9	57	38

Add or Subtract numbers in column

	a	b	c	d	e
# 1	57 +0	57 -0	37 +20	57 -20	17 +40
# 2	56 +1	57 -1	36 +21	57 -21	16 +41
# 3	55 +2	57 -2	35 +22	57 -22	15 +42
# 4	54 +3	57 -3	34 +23	57 -23	14 +43
# 5	53 +4	57 -4	33 +24	57 -24	13 +44
# 6	52 +5	57 -5	32 + 25	57 -25	12 +45
# 7	51 +6	57 -6	31 +26	57 -26	11 +46
# 8	50 +7	57 -7	30 +27	57 -27	10 +47
# 9	49 +8	57 -8	29 +28	57 -28	9 +48
# 10	48 +9	57 -9	28 +29	57 -29	8 +49

Summary 2 Chapter 2 Lesson 9
Number 58
5 tens – 1 ten = 4 tens   1 ten = 10 ones

a)	b)	c)	d)
1	4 18	1	4 18
29	5 8	19	58
+29	- 2 9	+39	- 19
58	2 9	58	39

Add or Subtract numbers in column

	a	b	c	d	e
# 1	58 +0	58 -0	38 +20	58 -20	18 +40
# 2	57 +1	58 -1	37 +21	58 -21	17 +41
# 3	56 +2	58 -2	36 +22	58 -22	16 +42
# 4	55 +3	58 -3	35 +23	58 -23	15 +43
# 5	54 +4	58 -4	34 +24	58 -24	14 +44
# 6	53 +5	58 -5	33 + 25	58 -25	13 +45
# 7	52 +6	58 -6	32 +26	58 -26	12 +46
# 8	51 +7	58 -7	31 +27	58 -27	11 +47
# 9	50 +8	58 -8	30 +28	58 -28	10 +48
# 10	49 +9	58 -9	29 +29	58 -29	9 +49

Summary 2 Chapter 2 Lesson 10
Number 59

a)	b)	c)	d)
29	59	19	59
+ 30	- 30	+40	-19
59	29	59	40

Add or Subtract numbers in column

	a	b	c	d	e
# 1	59 +0	59 -0	39 +20	59 -20	19 +40
# 2	58 +1	59 -1	38 +21	59 -21	18 +41
# 3	57 +2	59 -2	37 +22	59 -22	17 +42
# 4	56 +3	59 -3	36 +23	59 -23	16 +43
# 5	55 +4	59 -4	35 +24	59 -24	15 +44
# 6	54 +5	59 -5	34 + 25	59 -25	14 +45
# 7	53 +6	59 -6	33 +26	59 -26	13 +46
# 8	52 +7	59 -7	32 +27	59 -27	12 +47
# 9	51 +8	59 -8	31 +28	59 -28	11 +48
# 10	50 +9	59 -9	30 +29	59 -29	10 +49

Summary 2 Chapter 2 Lesson 11
Numbers from 50 to 54
Test 2 A
Add or Subtract

	a	b	c	d	e
# 1	49 + 1	50 - 1	29 +21	50 - 21	9 + 41
# 2	39 +11	50 - 11	19 + 31	50 - 31	50 - 41
# 3	48 + 3	51 - 3	28 +23	51 - 23	8 +43
# 4	38 + 13	51 -13	18 +33	51 -33	51 -43
# 5	47 +5	52 -5	27 +25	52 -25	7 +45
# 6	37 +15	52 -15	17 +35	52 -35	52 -45
# 7	46 + 7	53 - 7	26 +27	53 -27	6 +47
# 8	36 + 17	53 - 17	16 +37	53 -37	53 -47
# 9	45 +9	54 -9	25 +29	54 -29	7 +47
# 10	35 +19	54 -19	15 +39	54 -39	54 -49

Summary 2 Chapter 2 Lesson 11
Numbers from 5 to 59
Test 2 B

	a	b	c	d	e
# 1	45 +10	55 -10	25 +30	55 -30	55 -40
# 2	44 + 11	55 -11	24 + 31	55 -31	55 -41
# 3	44 +12	56 -12	24 +32	56 -32	56 -42
# 4	43 +13	56 -13	23 +33	56 -33	56 -43
# 5	43 +14	57 -14	23 +34	57 -34	57 -44
# 6	42 +15	57 -15	22 +35	57 -35	57 -45
# 7	42 +16	58 -16	22 +36	58 -36	58 -46
# 8	41 +17	58 -17	21 +37	58 -37	58 -47
# 9	41 +18	59 -18	21 +38	59 -38	59 -48
# 10	40 +19	59 -19	20 +39	59 -39	59 -49

Summary 2 Answers Chapter 2

Lesson 1 Number 50

	a	b	c	d	e
# 1	50,	50,	50,	30,	50
# 2	50,	49,	50,	29,	50
# 3	50,	48,	50,	28,	50
# 4	50,	47,	50,	27,	50
# 5	50,	46,	50,	26,	50
# 6	50,	45,	50,	25,	50
# 7	50,	44,	50,	24,	50
# 8	50,	43,	50,	23,	50
# 9	50,	42,	50,	22,	50
# 10	50,	41,	50,	21,	50

Lesson 2 Number 51

	a	b	c	d	e
# 1	51,	51,	51,	31,	51
# 2	51,	50,	51,	30,	51
# 3	51,	49,	51,	29,	51
# 4	51,	48,	51,	28,	51
# 5	51,	47,	51,	27,	51
# 6	51,	46,	51,	26,	51
# 7	51,	45,	51,	25,	51
# 8	51,	44,	51,	24,	51
# 9	51,	43,	51,	23,	51
# 10	51,	42,	51,	22,	51

Lesson 3 Number 52

	a	b	c	d	e
# 1	52,	52,	52,	32,	52
# 2	52,	51,	52,	31,	52
# 3	52,	50,	52,	30,	52
# 4	52,	49,	52,	29,	52
# 5	52,	48,	52,	28,	52
# 6	52,	47,	52,	27,	52
# 7	52,	46,	52,	26,	52
# 8	52,	45,	52,	25,	52
# 9	52,	44,	52,	24,	52
# 10	52,	43,	52,	23,	52

Summary 2 Answers Chapter 2

Lesson 4 Number 53

	a	b	c	d	e
# 1	53,	53,	53,	33,	53
# 2	53,	52,	53,	32,	53
# 3	53,	51,	53,	31,	53
# 4	53,	50,	53,	30,	53
# 5	53,	49,	53,	29,	53
# 6	53,	48,	53,	28,	53
# 7	53,	47,	53,	27,	53
# 8	53,	46,	53,	26,	53
# 9	53,	45,	53,	25,	53
# 10	53,	44,	53,	24,	53

Lesson 5 Number 54

	a	b	c	d	e
# 1	54,	54,	54,	34,	54
# 2	54,	53,	54,	33,	54
# 3	54,	52,	54,	32,	54
# 4	54,	51,	54,	31,	54
# 5	54,	50,	54,	30,	54
# 6	54,	49,	54,	29,	54
# 7	54,	48,	54,	28,	54
# 8	54,	47,	54,	27,	54
# 9	54,	46,	54,	26,	54
# 10	54,	45,	54,	25,	54

Lesson 6 Number 55

	a	b	c	d	e
# 1	55,	55,	55,	35,	55
# 2	55,	54,	55,	34,	55
# 3	55,	53,	55,	33,	55
# 4	55,	52,	55,	32,	55
# 5	55,	51,	55,	31,	55
# 6	55,	50,	55,	30,	55
# 7	55,	49,	55,	29,	55
# 8	55,	48,	55,	28,	55
# 9	55,	47,	55,	27,	55
# 10	55,	46,	55,	26,	55

Summary 2 Answers Chapter 2

Lesson 7 Number 56

	a	b	c	d	e
# 1	56,	56,	56,	36,	56
# 2	56,	55,	56,	35,	56
# 3	56,	54,	56,	34,	56
# 4	56,	53,	56,	33,	56
# 5	56,	52,	56,	32,	56
# 6	56,	51,	56,	31,	56
# 7	56,	50,	56,	30,	56
# 8	56,	49,	56,	29,	56
# 9	56,	48,	56,	28,	56
# 10	56,	47,	56,	27,	56

Lesson 8 Number 57

	a	b	c	d	e
# 1	57,	57,	57,	37,	57
# 2	57,	56,	57,	36,	57
# 3	57,	55,	57,	35,	57
# 4	57,	54,	57,	34,	57
# 5	57,	53,	57,	33,	57
# 6	57,	52,	57,	32,	57
# 7	57,	51,	57,	31,	57
# 8	57,	50,	57,	30,	57
# 9	57,	49,	57,	29,	57
# 10	57,	48,	57,	28,	57

Lesson 9 Number 58

	a	b	c	d	e
# 1	58,	58,	58,	38,	58
# 2	58,	57,	58,	37,	58
# 3	58,	56,	58,	36,	58
# 4	58,	55,	58,	35,	58
# 5	58,	54,	58,	34,	58
# 6	58,	53,	58,	33,	58
# 7	58,	52,	58,	32,	58
# 8	58,	51,	58,	31,	58
# 9	58,	50,	58,	30,	58
# 10	58,	49,	58,	29,	58

Summary 2 Answers Chapter 2

Lesson 10 Number 59

	a	b	c	d	e
# 1	59,	59,	59,	39,	59
# 2	59,	58,	59,	38,	59
# 3	59,	57,	59,	37,	59
# 4	59,	56,	59,	36,	59
# 5	59,	55,	59,	35,	59
# 6	59,	54,	59,	34,	59
# 7	59,	53,	59,	33,	59
# 8	59,	52,	59,	32,	59
# 9	59,	51,	59,	31,	59
# 10	59,	50,	59,	30,	59

Lesson 11 Test 2

Test 2 A

	a	b	c	d	e
# 1	50	49	50	29	50
# 2	50	39	50	19	9
# 3	51	48	51	28	51
# 4	51	38	51	18	8
# 5	52	47	52	27	52
# 6	52	37	52	17	7
# 7	53	46	53	26	53
# 8	53	36	53	16	6
# 9	54	45	54	25	54
# 10	54	35	54	15	5

Test 2 B

	a	b	c	d	e
# 1	55	45	55	25	15
# 2	55	44	55	24	14
# 3	56	44	56	24	14
# 4	56	43	56	23	13
# 5	57	43	57	23	13
# 6	57	42	57	22	12
# 7	58	42	58	22	12
# 8	58	41	58	21	11
# 9	59	41	59	21	11
# 10	59	40	59	20	10

Summary 2

Chapter 3 Addition or Subtraction Numbers (30-39)

Summary 2 Chapter 3 Contents
Addition or Subtraction Numbers (30-39)
Lesson 1 - 13
Basic Problems

Work with lesson 1, read rules and examples, 2) Solve the problems from contents, 3) Check answers in the end of chapter 1, 4) Work with the next lessons 2, 3, 4, 5, ..., 5) Solve Test 1 A, 6) Grading Scale: A =10 points, B=9points, C=8 points, D=7 points, F=0-6 points, 7) If you solve 10 problems correctly, you get an A, 8) If have an A, you can move on to next chapter, 9) If your grade is lower than an A, work with the mistakes, then solve Test 1B, 10) When you finish all of the chapters, solve Final Exam in last chapter.

Summary 2 Chapter 3 Lesson 1
Number 30
Add or Subtract numbers in column

3 tens – 1 ten = 2 tens   1 ten = 10 ones

a)	b)	c)	d)
1	2 10	1	2 10
19	3 0	18	30
+11	- 1 1	+12	- 12
30	1 9	30	18

Add or Subtract

	a	b	c	d	e
# 1	30 + 0	30 - 0	20 + 10	30 - 10	10 + 20
# 2	29 + 1	30 - 1	19 +11	30 - 11	9 + 21
# 3	28 + 2	30 - 2	18 + 12	30 - 12	8 + 22
# 4	27 + 3	30 - 3	17 + 13	30 - 13	7 + 23
# 5	26 + 4	30 - 4	16 + 14	30 - 14	6 + 24
# 6	25 + 5	30 - 5	15 + 15	30 - 15	5 + 25
# 7	24 + 6	30 - 6	14 + 16	30 - 16	4 + 26
# 8	23 + 7	30 - 7	13 + 17	30 - 17	3 + 27
# 9	22 + 8	30 - 8	12 + 18	30 - 18	2 + 28
# 10	21 + 9	30 - 9	11 + 19	30 - 19	1 + 29

Summary 2 Chapter 3 Lesson 2
Number 31
Add or Subtract numbers in column
3 tens – 1 ten = 2 tens   1 ten = 10 ones

a)	b)	c)	d)
1	2 11	1	2 11
19	3 1	18	31
+12	- 1 2	+13	- 18
31	1 9	31	13

Add or Subtract

	a	b	c	d	e
# 1	31	31	21	31	11
	+0	- 0	+10	-10	+20
# 2	30	31	20	31	10
	+1	-1	+11	-11	+ 21
# 3	29	31	19	31	9
	+ 2	- 2	+12	-12	+22
# 4	28	31	18	31	8
	+ 3	- 3	+13	- 13	+23
# 5	27	31	17	31	7
	+ 4	-4	+14	-14	+24
# 6	26	31	16	31	6
	+ 5	-5	+15	-15	+25
# 7	25	31	15	31	5
	+ 6	- 6	+ 16	- 16	+ 26
# 8	24	31	14	31	4
	+ 7	- 7	+17	-17	+ 27
# 9	23	31	13	31	3
	+ 8	- 8	+ 18	-28	+28
# 10	22	31	12	31	2
	+ 9	- 9	+19	-29	+29

Summary 2 Chapter 3 Lesson 3
Number 32

3 tens – 1 ten = 2 tens   1 ten = 10 ones

a)	b)	c)	d)
1	2 12	1	2 12
19	3 2	18	32
+13	- 1 3	+14	- 18
32	1 9	32	14

Add or Subtract numbers in column

	a	b	c	d	e
# 1	32 +0	32 -0	22 +10	32 -10	12 +20
# 2	31 +1	32 -1	21 +11	32 -11	11 +21
# 3	30 +2	32 -2	20 +12	32 - 12	10 +22
# 4	29 +3	32 -3	19 +13	32 -13	9 +23
# 5	28 +4	32 -4	18 +14	32 -14	8 +24
# 6	27 +5	32 -5	17 +15	32 -15	7 +25
# 7	26 +6	32 -6	16 +16	32 -16	6 +26
# 8	25 +7	32 - 7	15 +17	32 - 17	5 +27
# 9	24 +8	32 - 8	14 +18	32 -18	4 +28
# 10	23 +9	32 - 9	13 +19	32 -19	3 +29

Summary 2 Chapter 3 Lesson 4
Number 33

3 tens – 1 ten = 2 tens   1 ten = 10 ones

a)	b)	c)	d)
1	2 13	1	2 13
19	3 3	18	33
+14	- 1 4	+15	- 18
33	2 9	33	15

Add or Subtract numbers in column

	a	b	c	d	e
# 1	33	33	23	33	13
	+ 0	- 0	+10	-10	+20
# 2	32	33	22	33	12
	+ 1	-1	+11	-11	+21
# 3	31	33	21	33	11
	+ 2	- 2	+12	-12	+22
# 4	30	33	20	33	10
	+ 3	- 3	+13	-13	+23
# 5	29	33	19	33	9
	+ 4	- 4	+14	-14	+24
# 6	28	33	18	33	8
	+ 5	- 5	+15	-15	+25
# 7	27	33	17	33	7
	+ 6	- 6	+26	-16	+26
# 8	26	33	16	33	6
	+ 7	- 7	+17	-17	+27
# 9	25	33	15	33	5
	+ 8	- 8	+18	-18	+28
# 10	24	33	14	33	4
	+9	- 9	+19	-19	+29

Summary 2 Chapter 3 Lesson 5
Number 34
Add or Subtract numbers in column
3 tens – 1 ten = 2 tens   1 ten = 10 ones

a)	b)	c)	d)
1	2 14	1	2 14
19	3 4	18	34
+15	- 1 5	+15	- 19
34	1 9	34	15

Add or Subtract numbers in column

	a	b	c	d	e
# 1	34 +0	34 -0	24 +10	34 -10	14 +20
# 2	33 +1	34 -1	23 +11	34 -11	13 +21
# 3	32 +2	34 -2	22 +12	34 -12	12 +22
# 4	31 +3	34 -3	21 +13	34 -13	11 +23
# 5	30 +4	34 -4	20 +14	34 -14	10 +24
# 6	29 +5	34 -5	19 +15	34 -15	9 +25
# 7	28 +6	34 -6	18 +16	34 -16	8 +26
# 8	27 +7	34 -7	17 +17	34 -17	7 +27
# 9	26 +8	34 -8	16 +18	34 -18	6 +28
# 10	25 +9	34 -9	15 +19	34 -19	7 +27

Summary 2 Chapter 3 Lesson 6
Number 35
3 tens – 1 ten = 2 tens   1 ten = 10 ones

a)	b)	c)	d)
1	2 15	1	2 15
19	3 5	18	35
+16	- 1 6	+17	- 18
35	1 9	35	17

Add or Subtract numbers in column

	a	b	c	d	e
# 1	35	35	25	35	15
	+0	-0	+10	-10	+20
# 2	34	35	24	35	14
	+1	-1	+11	-11	+21
# 3	33	35	23	35	13
	+2	-2	+12	-12	+22
# 4	32	35	22	35	12
	+3	-3	+13	-13	+23
# 5	31	35	21	35	11
	+4	-4	+14	-14	+24
# 6	30	35	20	35	10
	+5	-5	+ 15	-15	+25
# 7	29	35	19	35	9
	+6	-6	+16	-16	+26
# 8	28	35	18	35	8
	+7	-7	+ 17	- 17	+27
# 9	27	35	17	35	7
	+8	-8	+18	-18	+28
# 10	26	35	16	35	6
	+9	- 9	+19	-19	+29

Summary 2 Chapter 3 Lesson 7
Number 36
Add or Subtract numbers in column
3 tens – 1 ten = 2 tens   1 ten = 10 ones

	a)	b)	c)	d)
	1	2 16	1	2 16
	19	3 6	18	36
	+17	- 1 7	+18	- 18
	36	1 9	36	18

Add or Subtract numbers in column

	a	b	c	d	e
# 1	36 +0	36 -0	26 +10	36 -10	16 +20
# 2	35 +1	36 -1	25 +11	36 -11	15 +21
# 3	34 +2	36 -2	24 +12	36 -12	14 +22
# 4	33 +3	36 -3	23 +13	36 -13	13 +23
# 5	32 +4	36 -4	22 +14	36 -14	12 +24
# 6	31 +5	36 -5	21 + 15	36 -15	11 +25
# 7	30 +6	36 -6	20 +16	36 -16	10 +26
# 8	29 +7	36 -7	19 +17	36 -17	9 +27
# 9	28 +8	36 -8	18 +18	36 -18	8 +28
# 10	27 +9	36 -9	17 +19	36 -19	7 + 29

Summary 2 Chapter 3 Lesson 8

Number 37

3 tens – 1 ten = 2 tens   1 ten = 10 ones

a)	b)	c)	d)
1	2 17	1	2 17
19	3 7	18	37
+18	- 19	+ 19	- 18
37	18	37	19

Add or Subtract numbers in column

	a	b	c	d	e
# 1	37 +0	37 -0	27 +10	37 -10	17 +20
# 2	36 +1	37 -1	26 +11	37 -11	16 +21
# 3	35 +2	37 -2	25 +12	37 -12	15 +22
# 4	34 +3	37 -3	24 +13	37 -13	14 +23
# 5	33 +4	37 -4	23 +14	37 -14	13 +24
# 6	32 +5	37 -5	22 + 15	37 -15	12 +25
# 7	31 +6	37 -6	21 +16	37 -16	11 +26
# 8	30 +7	37 -7	20 +17	37 -17	10 +27
# 9	29 +8	37 -8	19 +18	37 -18	9 +28
# 10	28 +9	37 -9	18 +19	37 -19	8 +29

Summary 2 Chapter 3 Lesson 9
Number 38
3 tens – 1 ten = 2 tens   1 ten = 10 ones

a)	b)	c)	d)
1	2 18	1	2 18
19	3 8	18	38
+19	- 1 9	+20	- 18
38	1 9	38	20

Add or Subtract numbers in column

	a	b	c	d	e
# 1	38 +0	38 -0	28 +10	38 -10	18 +20
# 2	37 +1	38 -1	27 +11	38 -11	17 +21
# 3	36 +2	38 -2	26 +12	38 -12	16 +22
# 4	35 +3	38 -3	25 +13	38 -13	15 +23
# 5	34 +4	38 -4	24 +14	38 -14	14 +24
# 6	33 +5	38 -5	23 + 15	38 -15	13 +25
# 7	32 +6	38 -6	22 +16	38 -16	12 +26
# 8	31 +7	38 -7	21 +17	38 -17	11 +27
# 9	30 +8	38 -8	20 +18	38 -18	10 +28
# 10	29 +9	38 -9	19 +19	38 -19	9 +29

Summary 2 Chapter 3 Lesson 10
Number 39

a)		b)		c)		d)	
	29		39		18		39
	+ 10		- 10		+21		-18
	39		19		39		21

Add or Subtract numbers in column

	a	b	c	d	e
# 1	39 +0	39 -0	29 +10	39 -20	19 +20
# 2	38 +1	39 -1	28 +11	39 -21	18 +21
# 3	37 +2	39 -2	27 +12	39 -22	17 +22
# 4	36 +3	39 -3	26 +13	39 -23	16 +23
# 5	35 +4	39 -4	25 +14	39 -24	15 +24
# 6	34 +5	39 -5	24 + 15	39 -25	14 +25
# 7	33 +6	39 -6	23 +16	39 -26	13 +26
# 8	32 +7	39 -7	22 +17	39 -27	12 +27
# 9	31 +8	39 -8	21 +18	39 -28	11 +28
# 10	30 +9	39 -9	20 +19	39 -29	10 +29

Summary 2 Chapter 3 Lesson 11
Numbers from 30 to 39
Test 3 A
Add or Subtract

	a	b
# 1	30 + 1	30 - 1
# 2	30 + 2	30 - 2
# 3	31 + 3	31 - 3
# 4	31 + 4	31 - 4
# 5	32 +5	32 -5
# 6	32 + 1	32 - 1
# 7	33 + 2	33 - 2
# 8	33 + 3	33 - 3
# 9	34 + 4	34 - 4
# 10	34 + 5	34 - 5

Summary 2 Chapter 3 Lesson 11
Numbers from 30 to 39
Test 3 B

	a	b
# 1	30 + 9	30 - 9
# 2	30 + 8	30 - 8
# 3	31 + 7	31 - 7
# 4	31 + 6	31 - 6
# 5	32 + 5	32 - 5
# 6	32 + 6	32 - 6
# 7	33 + 6	33 - 6
# 8	33 + 5	33 - 5
# 9	34 + 5	34 - 5
# 10	35 + 4	35 - 4

Summary 2 Chapter 3 Lesson 12
Review 2 A
Add or Subtract Numbers in Row
Numbers # 1-# 20 odd or # 1, # 3, # 5, # 7, # 9, # 11, # 13, # 15, # 17, # 19

 # 1 31 + 39

 # 2 39 + 32

 # 3 33 + 39

 # 4 39 + 34

 # 5 35 + 39

 # 6 70 – 31

 # 7 71 – 39

 # 8 72 – 33

 # 9 73 – 39

 # 10 74 – 35

Add or Subtract Numbers in Column

# 11	# 12	# 13	# 14	# 15
36	39	38	39	40
+39	+37	+ 39	+39	+ 39

# 16	# 17	# 18	# 19	# 20
75	76	77	78	79
-39	- 37	-39	-49	- 29

Summary 2 Chapter 3 Lesson 12
Review 2 B
Put right number instead question mark
Numbers # 1-# 20 odd or # 1, # 3, # 5, # 7, # 9, # 11, # 13, # 15, # 17, # 19

 # 1   41 + 39

 # 2   39 + 42

 # 3   43 + 39

 # 4   39 + 44

 # 5   35 + 49

 # 6   80 – 39

 # 7   81 - 32

 # 8   82 – 39

 # 9   83 - 34

 # 10   84 – 39

Add or Subtract Numbers in Column

# 11 36 +49	**# 12** 49 +37	**# 13** 48 + 39	**# 14** 39 +49	**# 15** 50 + 39
# 16 85 -39	**# 17** 86 -37	**# 18** 87 -39	**# 19** 88 -49	**# 20** 89 - 29

Summary 2 Chapter 3 Lesson 13
Final Exam 2 A
Add or Subtract Numbers in Row
Numbers # 1-# 20 odd or # 1, # 3, # 5, # 7, # 9, # 11, # 13, # 15, # 17, # 19

# 1	11 + 39
# 2	29 + 22
# 3	23 + 29
# 4	29 + 24
# 5	15 + 39
# 6	50 – 31
# 7	51 – 29
# 8	52 – 13
# 9	53 – 29
# 10	54 – 25

Add or Subtract Numbers in Column

# 11	# 12	# 13	# 14	# 15
29	29	28	19	40
+26	+27	+ 29	+39	+ 19

# 16	# 17	# 18	# 19	# 20
55	56	57	58	59
-28	-38	-18	-39	- 19

Summary 2 Chapter 3 Lesson 13
Final Exam 2 B
Put right number instead question mark
Numbers # 1 - # 20 odd or # 1, # 3, # 5, # 7, # 9, # 11, # 13, # 15, # 17, # 19

 # 1 29 + 31

 # 2 32 + 29

 # 3 39 + 23

 # 4 34 + 29

 # 5 29 + 35

 # 6 60 – 31

 # 7 61 - 22

 # 8 62 – 13

 # 9 63 - 24

 # 10 64 – 29

Add or Subtract Numbers in Column

# 11 29 +36	# 12 29 +37	# 13 38 + 29	# 14 29 +39	# 15 40 + 29
# 16 65 -26	# 17 66 -37	# 18 67 -29	# 19 68 -29	# 20 69 - 29

Summary 2 Answers Chapter 3

Lesson 1 Number 30

	a	b	c	d	e
# 1	30,	30,	30,	20,	30
# 2	30,	29,	30,	19,	30
# 3	30,	28,	30,	18,	30
# 4	30,	27,	30,	17,	30
# 5	30,	26,	30,	16,	30
# 6	30,	25,	30,	15,	30
# 7	30,	24,	30,	14,	30
# 8	30,	23,	30,	13,	30
# 9	30,	22,	30,	12,	30
# 10	30,	21,	30,	11,	30

Lesson 2 Number 31

	a	b	c	d	e
# 1	31,	31,	31,	21,	31
# 2	31,	30,	31,	20,	31
# 3	31,	29,	31,	19,	31
# 4	31,	28,	31,	18,	31
# 5	31,	27,	31,	17,	31
# 6	31,	26,	31,	16,	31
# 7	31,	25,	31,	15,	31
# 8	31,	24,	31,	14,	31
# 9	31,	23,	31,	13,	31
# 10	31,	22,	31,	12,	31

Lesson 3 Number 32

	a	b	c	d	e
# 1	32,	32,	32,	22,	32
# 2	32,	31,	32,	21,	32
# 3	32,	30,	32,	20,	32
# 4	32,	29,	32,	19,	32
# 5	32,	28,	32,	18,	32
# 6	32,	27,	32,	17,	32
# 7	32,	26,	32,	16,	32
# 8	32,	25,	32,	15,	32
# 9	32,	24,	32,	14,	32
# 10	32,	23,	32,	13,	32

Summary 2 Answers Chapter 3

Lesson 4 Number 33

	a, b, c, d, e
# 1	33, 33, 33, 23, 33
# 2	33, 32, 33, 22, 33
# 3	33, 31, 33, 21, 33
# 4	33, 30, 33, 20, 33
# 5	33, 29, 33, 19, 33
# 6	33, 28, 33, 18, 33
# 7	33, 27, 33, 17, 33
# 8	33, 26, 33, 16, 33
# 9	33, 25, 33, 15, 33
# 10	33, 24, 33, 14, 33

Lesson 5 Number 34

	a, b, c, d, e
# 1	34, 34, 34, 24, 34
# 2	34, 33, 34, 23, 34
# 3	34, 32, 34, 22, 34
# 4	34, 31, 34, 21, 34
# 5	34, 30, 34, 20, 34
# 6	34, 29, 34, 19, 34
# 7	34, 28, 34, 18, 34
# 8	34, 27, 34, 17, 34
# 9	34, 26, 34, 16, 34
# 10	34, 25, 34, 15, 34

Lesson 6 Number 35

	a, b, c, d, e
# 1	35, 35, 35, 25, 35
# 2	35, 34, 35, 24, 35
# 3	35, 33, 35, 23, 35
# 4	35, 32, 35, 22, 35
# 5	35, 31, 35, 21, 35
# 6	35, 30, 35, 20, 35
# 7	35, 29, 35, 19, 35
# 8	35, 28, 35, 18, 35
# 9	35, 27, 35, 17, 35
# 10	35, 26, 35, 16, 35

Summary 2 Answers Chapter 3

Lesson 7 Number 36

	a	b	c	d	e
# 1	36,	36,	36,	26,	36
# 2	36,	35,	36,	25,	36
# 3	36,	34,	36,	24,	36
# 4	36,	33,	36,	23,	36
# 5	36,	32,	36,	22,	36
# 6	36,	31,	36,	21,	36
# 7	36,	30,	36,	20,	36
# 8	36,	29,	36,	19,	36
# 9	36,	28,	36,	18,	36
# 10	36,	27,	36,	17,	36

Lesson 8 Number 37

	a	b	c	d	e
# 1	37,	37,	37,	27,	37
# 2	37,	36,	37,	26,	37
# 3	37,	35,	37,	25,	37
# 4	37,	34,	37,	24,	37
# 5	37,	33,	37,	23,	37
# 6	37,	32,	37,	22,	37
# 7	37,	31,	37,	21,	37
# 8	37,	30,	37,	20,	37
# 9	37,	29,	37,	19,	37
# 10	37,	28,	37,	18,	37

Lesson 9 Number 38

	a	b	c	d	e
# 1	38,	38,	38,	28,	38
# 2	38,	37,	38,	27,	38
# 3	38,	36,	38,	26,	38
# 4	38,	35,	38,	25,	38
# 5	38,	34,	38,	24,	38
# 6	38,	33,	38,	23,	38
# 7	38,	32,	38,	22,	38
# 8	38,	31,	38,	21,	38
# 9	38,	30,	38,	20,	38
# 10	38,	29,	38,	19,	38

Lesson 10 Number 39

	a	b	c	d	e
# 1	39,	39,	39,	19,	39
# 2	39,	38,	39,	18,	39
# 3	39,	37,	39,	17,	39
# 4	39,	36,	39,	16,	39
# 5	39,	35,	39,	15,	39
# 6	39,	34,	39,	14,	39
# 7	39,	33,	39,	13,	39
# 8	39,	32,	39,	12,	39
# 9	39,	31,	39,	11,	39
# 10	39,	30,	39,	10,	39

Lesson 11 Test 3

Test 3 A

	a	b
# 1	31	29
# 2	32	28
# 3	34	28
# 4	35	27
# 5	37	27
# 6	33	31
# 7	35	31
# 8	36	30
# 9	38	30
# 10	39	29

Test 3 B

	a	b
# 1	39	21
# 2	38	22
# 3	38	24
# 4	37	25
# 5	37	27
# 6	38	26
# 7	39	27
# 8	38	28
# 9	39	29
# 10	39	31

Summary 2 Answers Chapter 3

Lesson 12 Review 2 A

# 1	70	# 5	74	# 9	34	# 13	77	# 17	39
# 2	71	# 6	39	# 10	39	# 14	78	# 18	38
# 3	72	# 7	32	# 11	75	# 15	79	# 19	29
# 4	73	# 8	39	# 12	76	# 16	36	# 20	50

Review 2 B

# 1	80	# 5	84	# 9	49	# 13	87	# 17	49
# 2	81	# 6	41	# 10	45	# 14	88	# 18	48
# 3	82	# 7	49	# 11	85	# 15	89	# 19	39
# 4	83	# 8	43	# 12	86	# 16	46	# 20	60

Lesson 13 Final Exam 2 A

# 1	50	# 5	54	# 9	24	# 13	57	# 17	18
# 2	51	# 6	19	# 10	29	# 14	58	# 18	39
# 3	52	# 7	22	# 11	55	# 15	59	# 19	19
# 4	53	# 8	39	# 12	56	# 16	27	# 20	40

Final Exam 2 B

# 1	60	# 5	64	# 9	39	# 13	67	# 17	29
# 2	61	# 6	29	# 10	35	# 14	68	# 18	38
# 3	62	# 7	39	# 11	65	# 15	69	# 19	39
# 4	63	# 8	49	# 12	66	# 16	39	# 20	40

Math Help

for College

Summary 3

Summary 3 Pretest 3
(25 minutes)

Name
Read directions carefully. Answers must be in the indicated place.
Show all steps for full credits. Each problem is worth 5 points.
Do only 1 operation per 1 step. Solve 1 problem 5 minutes

1) Multiply $\begin{array}{r} 605 \\ \underline{\times 403} \end{array}$

1)______

2) Divide $182508 \div 201$

2)____

3) Evaluate $180180 \div 90 \times 2$

3)______

4) Evaluate $480480 - 360360 \div 6 \times 5$

4)______

5) Evaluate $150 \div 5 \times 2 - 60 \div 5 \times 3$

5)______

Summary 3

Chapter 1 Addition or Subtraction Numbers (60-69)

Summary 3 Chapter 1 Contents
Numbers 60 – 69
Lesson 1 - 11
Basic Problems

Work with lesson 1, read rules and examples, 2) Solve the problems from contents, 3) Check answers in the end of chapter 1, 4) Work with the next lessons 2, 3, 4, 5, ..., 5) Solve Test 1 A, 6) Grading Scale: A =10 points, B=9points, C=8 points, D=7 points, F=0-6 points, 7) If you solve 10 problems correctly, you get an A, 8) If have an A, you can move on to next chapter, 9) If your grade is lower than an A, work with the mistakes, then solve Test 1B, 10) When you finish all of the chapters, solve Final Exam in last chapter. Lesson = L

Summary 3 Chapter 1 Lesson 1
Review 1 A
Add or Subtract numbers in row

Add

1) 9 + 1
2) 8 + 2
3) 7 + 3
4) 6 + 4
5) 5 + 5
6) 9 + 2
7) 8 + 3
8) 7 + 4
9) 6 + 5
10) 9 + 3
11) 8 + 4
12) 7 + 5
13) 6 + 6
14) 9 + 4
15) 8 + 5
16) 7 + 6
17) 9 + 5
18) 8 + 6
19) 7 + 7
20) 9 + 6
21) 8 + 7
22) 9 + 7
23) 8 + 8
24) 9 + 8
25) 9 + 9

Subtract

1) 10 – 1
2) 10 – 2
3) 10 – 3
4) 10 – 4
5) 10 – 5
6) 11 - 2
7) 11 – 3
8) 11 – 4
9) 11 – 5
10) 12 – 3
11) 12 – 4
12) 12 – 5
13) 12 – 6
14) 13 – 4
15) 13 – 5
16) 13 – 6
17) 14 – 5
18) 14 – 6
19) 14 – 7
20) 15 – 6
21) 15 – 7
22) 16 – 7
23) 16 – 8
24) 17 – 8
25) 18 - 9

Answers Lesson 1 Review 1 A

Add1)10 2)103)104)105)106)117)118)119)1110)1211)1212)1213)1214)13 15)13 16) 13 17) 14 18) 14 19) 14 20) 15 21) 15 22) 16 23) 16 24) 17 25) 18 Subtract 1) 9 2) 83) 7 4) 6 5) 5 6) 9 7) 8 8) 7 9) 6 10) 9 11) 8 12) 7 13) 6 14) 9 15) 8 16) 7 17) 9 18) 8 19) 7 20) 9 21) 8 22) 9 23) 8 24) 9 25) 9

Summary 3 Chapter 1 Lesson 1
Review 1 B
Add or Subtract numbers in row

Number 20
6 + 14 = 14 + 6 =10+4+6=10+10= 20
20 – 14 = 20 – 10 – 4 = 10 – 4 = 6
Number 21
7 + 14 = 14 + 7 = 10 + 4 + 7 =10 + 11 = 21
21 – 14 = 21 – 11 – 3 = 10 – 3 = 7
Number 22
4 + 18 = 18 + 4 = 10 + 8 + 4 = 10 + 12 = 22
22 – 14 = 22 – 12 – 2 = 10 – 2 = 8
Number 23
9 + 14 = 14 + 9 = 10 + 4 + 9=10+13 = 23
23 – 14 = 23 – 13 – 1 = 10 – 1 = 9
Number 24
10 + 14 = 10 + 10 + 4 = 20 + 4 = 24
24 – 14 = 24 – 10 – 4 = 14 – 4 = 10
Number 25
11 + 14 = 10 + 1 + 10 + 4 = 20 + 5 = 25
25 – 14 = 25 – 10 – 4 = 15 – 4 = 11
Numbers 26
12 + 14 = 10 + 2 + 10 + 4 = 20 + 6 = 26
26 - 14 = 26 – 10 – 4 = 16 – 4 = 12
Number 27
13 + 14 = 10 + 3 + 10 + 4 = 20 + 7 = 27
27 – 14 = 27 – 10 – 4 = 17 – 4 = 13
Number 28
14 + 14 = 10 + 4 + 10 + 4 = 20 + 8 = 28
28 – 14 = 28 – 10 – 4 = 18 – 4 = 14
Number 29
15 + 14 = 10 + 5 + 10 + 4 = 20 + 9 = 29
29 – 14 = 29 – 10 – 4 = 19 – 4 = 15

Summary 3 Chapter 1 Lesson 2
Review 1 A
Add or Subtract numbers in column

5 tens – 1 ten = 4 tens **10 ones = 1 ten**

1) Number 50

1	4 10	1	4 10
a) 29	b) 5 0	c) 19	d) 50
+21	- 2 1	+31	- 19
50	2 9	50	31

2) Number 51

1	4 11	1	4 11
a) 29	b) 5 1	c) 19	d) 51
+22	- 2 2	+32	- 19
51	2 9	51	32

3) Number 52

1	4 12	1	4 12
a) 29	b) 5 2	c) 19	d) 52
+23	- 2 3	+33	- 19
52	2 9	52	33

4) Number 53

1	4 13	1	4 13
a) 29	b) 5 3	c) 19	d) 53
+24	- 2 4	+34	- 19
53	2 9	53	34

5) Numbers 54

1	4 14	1	4 14
1) 29	2) 5 4	3) 19	4) 54
+25	- 2 5	+ 35	- 19
54	2 9	54	35

Summary 3 Chapter 1 Lesson 2
Review 1 B
Add or Subtract numbers in column

5 tens – 1 ten = 4 tens **10 ones = 1 ten**

6) Numbers 55

a)	b)	c)	d)
1	4 15	1	4 15
29	5 5	19	55
+26	- 2 6	+36	- 19
55	2 9	55	36

7) Numbers 56

a)	b)	c)	d)
1	4 16	1	4 16
29	5 6	19	56
+27	- 2 7	+37	- 19
56	2 9	56	37

8) Numbers 57

a)	b)	c)	d)
1	4 17	1	4 17
29	5 7	19	57
+28	- 2 8	+38	- 19
57	2 9	57	38

9) Numbers 58

a)	b)	c)	d)
1	4 18	1	4 18
29	5 8	19	58
+29	- 2 9	+39	- 19
58	2 9	58	39

10) Numbers 59

a)	b)	c)	d)
29	59	19	59
+ 30	- 30	+40	-19
59	29	59	40

Summary 3 Chapter 1 Lesson 3
Review 1 A
Add or Subtract Numbers in Row

20 – 16 = 20 – 10 – 6 = 10 – 6 = 4

# 1	31 + 19
# 2	29 + 22
# 3	13 + 39
# 4	29 + 24
# 5	25 + 29
# 6	20 – 11
# 7	20 – 12
# 8	20 – 13
# 9	20 – 14
# 10	20 – 15

Add or Subtract Numbers in Column

# 11 26 +29	# 12 29 +27	# 13 28 + 29	# 14 19 +39	# 15 40 + 19
# 16 55 -29	# 17 56 -27	# 18 57 -29	# 19 58 -39	# 20 59 - 19

Answers Lesson 3 Review 1 A

# 1	50	# 11	55
# 2	51	# 12	56
# 3	52	# 13	57
# 4	53	# 14	58
# 5	54	# 15	59
# 6	9	# 16	26
# 7	8	# 17	29
# 8	7	# 18	28
# 9	6	# 19	19
# 10	5	# 20	40

Summary 3 Chapter 1 Lesson 3
Review 1 B
Add or Subtract Numbers in Row

30 – 16 = 30 – 10 – 6 = 20 – 6 = 14

# 1	11 + 39
# 2	29 + 22
# 3	23 + 29
# 4	29 + 34
# 5	15 + 39
# 6	30 – 11
# 7	30 – 12
# 8	30 – 13
# 9	30 – 14
# 10	30 – 15

Add or Subtract Numbers in Column

# 11	# 12	# 13	# 14	# 15
29	27	29	39	19
+26	+29	+ 27	+19	+ 40

# 16	# 17	# 18	# 19	# 20
55	56	57	58	59
-26	-29	-18	-39	- 19

Answers Lesson 3 Review 1 B

# 1	50	# 11	55
# 2	51	# 12	56
# 3	52	# 13	56
# 4	63	# 14	58
# 5	54	# 15	59
# 6	19	# 16	29
# 7	18	# 17	27
# 8	17	# 18	39
# 9	16	# 19	19
# 10	15	# 20	40

Summary 3 Chapter 1 Lesson 4
Add and Subtract three numbers
Three numbers add in any order

11 + 14 + 19 = 11 + 19 + 14 = 30 + 14 = 44

11 + 14 + 19 = 25 + 19 = 44

11 + 14 + 19 = 11 + 33 = 44

Review 1 A
Add three numbers

	a	b	c	d
# 1	20 + 0 + 9	30 + 0 + 9	40 + 0 + 9	50 + 0 + 9
# 2	21 + 1 + 8	31 + 1 + 8	41 + 1 + 8	51 + 1 + 8
# 3	22 + 2 + 7	32 + 2 + 7	42 + 2 + 7	52 + 2 + 7
# 4	23 + 3 + 6	33 + 3 + 6	43 + 3 + 6	53 + 3 + 6
# 5	24 + 4 + 5	34 + 4 + 5	44 + 4 + 5	54 + 4 + 5
# 6	25 + 5 + 4	35 + 5 + 4	45 + 5 + 4	55 + 5 + 4
# 7	26 + 6 + 3	36 + 6 + 3	46 + 6 + 3	56 + 6 + 3
# 8	27 + 7 + 2	37 + 7 + 2	47 + 7 + 2	57 + 7 + 2
# 9	28 + 8 + 1	38 + 8 + 1	48 + 8 + 1	58 + 8 + 1
# 10	29 + 9 + 0	39 + 9 + 0	49 + 9 + 0	59 + 9 + 0

Answers Lesson 4 Review 1 A

	a	b	c	d
# 1	29	39	49	59
# 2	30	40	50	60
# 3	31	41	51	61
# 4	32	42	52	62
# 5	33	43	53	63
# 6	34	44	54	64
# 7	35	45	55	65
# 8	36	46	56	66
# 9	37	47	57	67
# 10	38	48	58	68

Summary 3 Chapter 1 Lesson 4
Add and Subtract three numbers
To subtract three numbers, work from left to right

19 – 1 – 9 = 18 – 9 = 9

18 – 2 – 8 = 16 – 8 = 8

18 – 8 – 2 = 10 – 2 = 8

Review 1 B
Subtract three numbers

	a	b	c	d
# 11	60 – 9 – 0	70 – 9 – 0	80 – 9 – 0	90 – 9 – 0
# 12	61 – 8 – 1	71 – 8 – 1	81 – 8 – 1	91 – 8 – 1
# 13	62 – 7 – 2	72 – 7 – 2	82 - 7 – 2	92 – 7 – 2
# 14	63 – 6 – 3	73 – 6 – 3	83 – 6 – 3	93 – 6 – 3
# 15	64 – 5 – 4	74 – 5 – 4	84 – 5 – 4	94 – 5 – 4
# 16	65 – 4 – 5	75 – 4 - 5	85 - 4 – 5	95 – 4 – 5
# 17	66 – 3 – 6	76 – 3 – 6	86 – 3 – 6	96 – 3 – 6
# 18	67 – 2 – 7	77 – 2 – 7	87 – 2 – 7	97 – 2 – 7
# 19	68 – 1 – 8	78 – 1 – 8	88 – 1 – 8	98 – 1 – 8
# 20	69 – 0 – 9	79 – 0 – 9	89 – 0 – 9	99 – 0 - 9

Answers Lesson 4 Review 1 B

	a	b	c	d
# 11	51	61	71	81
# 12	52	62	72	82
# 13	53	63	73	83
# 14	54	64	74	84
# 15	55	65	75	85
# 16	56	66	76	86
# 17	57	67	77	87
# 18	58	68	78	88
# 19	59	69	79	89
# 20	60	70	80	90

Summary 3 Chapter 1 Lesson 5
Two operations with numbers
To do two operations work from left to right

19 + 9 - 1 = 26 - 1 = 25

18 + 8 - 2 = 26 - 2 = 24

Review 1 A
Add or Subtract numbers

	a	b	c	d
# 1	20 + 9 – 0	30 + 9 – 0	40 + 9 – 0	50 + 9 - 0
# 2	21 + 8 – 1	31 + 8 – 1	41 + 8 – 1	51 + 8 - 1
# 3	22 + 7 – 2	32 + 7 – 2	42 + 7 – 2	52 + 7 - 2
# 4	23 + 6 – 3	33 + 6 – 3	43 + 6 – 3	53 + 6 - 3
# 5	24 + 5 – 4	34 + 5 – 4	44 + 5 – 4	54 + 5 - 4
# 6	25 + 4 – 5	35 + 4 – 5	45 + 4 – 5	55 + 4 - 5
# 7	26 + 3 – 6	36 + 3 – 6	46 + 3 – 6	56 + 3 - 6
# 8	27 + 2 – 7	37 + 2 – 7	47 + 2 – 7	57 + 2 - 7
# 9	28 + 1 – 8	38 + 1 – 8	48 + 1 – 8	58 + 1 - 8
# 10	29 + 0 – 9	39 + 0 – 9	49 + 0 – 9	59 + 0 - 9

Answers Lesson 5 Review 1 A

	a	b	c	d
# 1	29	39	49	59
# 2	28	38	48	58
# 3	27	37	47	57
# 4	26	36	46	56
# 5	25	35	45	55
# 6	24	34	44	54
# 7	23	33	43	53
# 8	22	32	42	52
# 9	21	31	41	51
# 10	20	30	40	50

Summary 3 Chapter 1 Lesson 5

Two operations with numbers

To do two operations work from left to right

19 - 8 + 9 = 11 + 9 = 20

18 - 9 + 8 = 9 + 8 = 17

Review 1 B

Subtract or Add Numbers

	a	b	c	d
# 11	20 - 9 + 0	30 - 9 + 0	40 - 9 + 0	50 - 9 + 0
# 12	21 - 8 + 1	31 - 8 + 1	41 - 8 + 1	51 - 8 + 1
# 13	22 - 7 + 2	32 - 7 + 2	42 - 7 + 2	52 - 7 + 2
# 14	23 - 6 + 3	33 - 6 + 3	43 - 6 + 3	53 - 6 + 3
# 15	24 - 5 + 4	34 - 5 + 4	44 - 5 + 4	54 - 5 + 4
# 16	25 - 4 + 5	35 - 4 + 5	45 - 4 + 5	55 - 4 + 5
# 17	26 - 3 + 6	36 - 3 + 6	46 - 3 + 6	56 - 3 + 6
# 18	27 - 2 + 7	37 - 2 + 7	47 - 2 + 7	57 - 2 + 7
# 19	28 - 1 + 8	38 - 1 + 8	48 - 1 + 8	58 - 1 + 8
# 20	29 - 0 + 9	39 - 0 + 9	49 - 0 + 9	59 - 0 + 9

Answers Lesson 5 Review 1 B

	a	b	c	d
# 11	11	21	31	41
# 12	14	24	34	44
# 13	17	27	37	47
# 14	20	30	40	50
# 15	23	33	43	53
# 16	26	36	46	56
# 17	29	39	49	59
# 18	32	42	52	62
# 19	35	45	55	65
# 20	38	48	58	68

Summary 3 Chapter 1 Lesson 6
Number 60

31 + 29 =30 + 1 + 20 + 9 = 30 + 20+ 1 + 9 = 50 + 10 = 60

60 – 29 = 60 – 20 – 9 = 40 – 9 = 31

22 + 38 = 20 + 2 + 30 + 8 = 20 + 30 + 2 + 8 = 50 + 10 = 60

60 – 38 = 60 – 30 – 8 = 30 – 8 = 22

Add or Subtract numbers in row

	a	b	c	d
# 1	60 – 0	40 + 20	60 – 20	60 - 40
# 2	60 – 1	39 + 21	60 – 21	60 - 41
# 3	60 – 2	38 + 22	60 – 22	60 - 42
# 4	60 – 3	37 + 23	60 – 23	60 - 43
# 5	60 – 4	36 + 24	60 – 24	60 - 44
# 6	60 – 5	35 + 25	60 – 25	60 - 45
# 7	60 – 6	34 + 26	60 – 26	60 - 46
# 8	60 – 7	33 + 27	60 – 27	60 - 47
# 9	60 – 8	32 + 28	60 – 28	60 - 48
# 10	60 – 9	31 + 29	60 – 29	60 - 49
# 11	60 – 10	30 + 30	60 – 30	60 – 50
# 12	60 – 11	29 + 31	60 – 31	60 - 51
# 13	60 – 12	28 + 32	60 – 32	60 – 52
# 14	60 – 13	27 + 33	60 – 33	60 - 53
# 15	60 – 14	26 + 34	60 – 34	60 – 54
# 16	60 – 15	25 + 35	60 – 35	60 - 55
# 17	60 – 16	24 + 36	60 – 36	60 – 56
# 18	60 – 17	23 + 37	60 – 37	60 – 57
# 19	60 – 18	22 + 38	60 – 38	60 - 58
# 20	60 – 19	21 + 39	60 – 39	60 – 59

Summary 3 Chapter 1 Lesson 6
Number 61

32 + 29 =30 + 2 + 20 + 9 = 30 + 20 + 2 + 9 = 50 + 11 = 61
61 – 29 = 61 – 20 – 9 = 41 – 9 = 32
22 + 39 = 20 + 2 + 30 + 9 = 20 + 30 + 2 + 9 = 50 + 11 = 61
61 – 39 = 61 – 30 – 9 = 31 – 9 = 22

Add or Subtract numbers in row

	a	b	c	d
# 1	61 – 0	41 + 20	61 – 20	61 - 40
# 2	61 – 1	40 + 21	61 – 21	61 - 41
# 3	61 – 2	39 + 22	61 – 22	61 - 42
# 4	61 – 3	38 + 23	61 – 23	61 - 43
# 5	61 – 4	37 + 24	61 – 24	61 - 44
# 6	61 – 5	36 + 25	61 – 25	61 - 45
# 7	61 – 6	35 + 26	61 – 26	61 - 46
# 8	61 – 7	34 + 27	61 – 27	61 - 47
# 9	61 – 8	33 + 28	61 – 28	61 - 48
# 10	61 – 9	32 + 29	61 – 29	61 - 49
# 11	61 – 10	31 + 30	61 – 30	61 – 50
# 12	61 – 11	30 + 31	61 – 31	61 - 51
# 13	61 – 12	29 + 32	61 – 32	61 – 52
# 14	61 – 13	28 + 33	61 – 33	61 - 53
# 15	61 – 14	27 + 34	61 – 34	61 – 54
# 16	61 – 15	26 + 35	61 – 35	61 - 55
# 17	61 – 16	25 + 36	61 – 36	61 – 56
# 18	61 – 17	24 + 37	61 – 37	61 – 57
# 19	61 – 18	23 + 38	61 – 38	61 - 58
# 20	61 – 19	22 + 39	61 – 39	61 – 59

Summary 3 Chapter 1 Lesson 7
Number 62

33 + 29 =30 + 3 + 20 + 9 = 30 + 20 + 3 + 9 = 50 + 12 = 62

62 – 29 = 62 – 20 – 9 = 42 – 9 = 33

23 + 39 = 20 + 3 + 30 + 9 = 20 + 30 + 3 + 9 = 50 + 12 = 62

62 – 39 = 62 – 30 – 9 = 32 – 9 = 23

Add or Subtract numbers in row

	a	b	c	d
# 1	62 – 0	42 + 20	62 – 20	62 - 40
# 2	62 – 1	41 + 21	62 – 21	62 - 41
# 3	62 – 2	40 + 22	62 – 22	62 - 42
# 4	62 – 3	39 + 23	62 – 23	62 - 43
# 5	62 – 4	38 + 24	62 – 24	62 - 44
# 6	62 – 5	37 + 25	62 – 25	62 - 45
# 7	62 – 6	36 + 26	62 – 26	62 - 46
# 8	62 – 7	35 + 27	62 – 27	62 - 47
# 9	62 – 8	34 + 28	62 – 28	62 - 48
# 10	62 – 9	33 + 29	62 – 29	62 - 49
# 11	62 – 10	32 + 30	62 – 30	62 – 50
# 12	62 – 11	31 + 31	62 – 31	62 - 51
# 13	62 – 12	30 + 32	62 – 32	62 – 52
# 14	62 – 13	29 + 33	62 – 33	62 - 53
# 15	62 – 14	28 + 34	62 – 34	62 – 54
# 16	62 – 15	27 + 35	62 – 35	62 - 55
# 17	62 – 16	26 + 36	62 – 36	62 – 56
# 18	62 – 17	25 + 37	62 – 37	62 – 57
# 19	62 – 18	24 + 38	62 – 38	62 - 58
# 20	62 – 19	23 + 39	62 – 39	62 – 59

Summary 3 Chapter 1 Lesson 7
Number 63

34 + 29 =30 + 4 + 20 + 9 = 30 + 20 + 4 + 9 = 50 + 13 = 63

63 – 29 = 63 – 20 – 9 = 43 – 9 = 34

24 + 39 = 20 + 4 + 30 + 9 = 20 + 30 + 4 + 9 = 50 + 13 = 63

63 – 39 = 63 – 30 – 9 = 33 – 9 = 24

Add or Subtract numbers in row

	a	b	c	d
# 1	63 – 0	43 + 20	63 – 20	63 - 40
# 2	63 – 1	42 + 21	63 – 21	63 - 41
# 3	63 – 2	41 + 22	63 – 22	63 - 42
# 4	63 – 3	40 + 23	63 – 23	63 - 43
# 5	63 – 4	39 + 24	63 – 24	63 - 44
# 6	63 – 5	38 + 25	63 – 25	63 - 45
# 7	63 – 6	37 + 26	63 – 26	63 - 46
# 8	63 – 7	36 + 27	63 – 27	63 - 47
# 9	63 – 8	35 + 28	63 – 28	63 - 48
# 10	63 – 9	34 + 29	63 – 29	63 - 49
# 11	63 – 10	33 + 30	63 – 30	63 – 50
# 12	63 – 11	32 + 31	63 – 31	63 - 51
# 13	63 – 12	31 + 32	63 – 32	63 – 52
# 14	63 – 13	30 + 33	63 – 33	63 - 53
# 15	63 – 14	29 + 34	63 – 34	63 – 54
# 16	63 – 15	28 + 35	63 – 35	63 - 55
# 17	63 – 16	27 + 36	63 – 36	63 – 56
# 18	63 – 17	26 + 37	63 – 37	63 – 57
# 19	63 – 18	25 + 38	63 – 38	63 - 58
# 20	63 – 19	24 + 39	63 – 39	63 – 59

Summary 3 Chapter 1 Lesson 8

Number 64

35 + 29 =30 + 5 + 20 + 9 = 30 + 20 + 5 + 9 = 50 + 14 = 64

64 – 29 = 64 – 20 – 9 = 44 – 9 = 35

25 + 39 = 20 + 5 + 30 + 9 = 20 + 30 + 5 + 9 = 50 + 14 = 64

64 – 39 = 64 – 30 – 9 = 34 – 9 = 25

Add or Subtract numbers in row

	a	b	c	d
# 1	64 – 0	44 + 20	64 – 20	64 - 40
# 2	64 – 1	43 + 21	64 – 21	64 - 41
# 3	64 – 2	42 + 22	64 – 22	64 - 42
# 4	64 – 3	41 + 23	64 – 23	64 - 43
# 5	64 – 4	40 + 24	64 – 24	64 - 44
# 6	64 – 5	39 + 25	64 – 25	64 - 45
# 7	64 – 6	38 + 26	64 – 26	64 - 46
# 8	64 – 7	37 + 27	64 – 27	64 - 47
# 9	64 – 8	36 + 28	64 – 28	64 - 48
# 10	64 – 9	35 + 29	64 – 29	64 - 49
# 11	64 – 10	34 + 30	64 – 30	64 – 50
# 12	64 – 11	33 + 31	64 – 31	64 - 51
# 13	64 – 12	32 + 32	64 – 32	64 – 52
# 14	64 – 13	31 + 33	64 – 33	64 - 53
# 15	64 – 14	30 + 34	64 – 34	64 – 54
# 16	64 – 15	29 + 35	64 – 35	64 - 55
# 17	64 – 16	28 + 36	64 – 36	64 – 56
# 18	64 – 17	27 + 37	64 – 37	64 – 57
# 19	64 – 18	26 + 38	64 – 38	64 - 58
# 20	64 – 19	25 + 39	64 – 39	64 – 59

Summary 3 Chapter 1 Lesson 8
Number 65

36 + 29 =30 + 6 + 20 + 9 = 30 + 20 + 6 + 9 = 50 + 15 = 65

65 – 29 = 65 – 20 – 9 = 45 – 9 = 36

26 + 39 = 20 + 6 + 30 + 9 = 20 + 30 + 6 + 9 = 50 + 15 = 65

65 – 39 = 65 – 30 – 9 = 35 – 9 = 26

Add or Subtract numbers in row

	a	b	c	d
# 1	65 – 0	45 + 20	65 – 20	65 - 40
# 2	65 - 1	44 + 21	65 – 21	65 - 41
# 3	65 – 2	43 + 22	65 – 22	65 - 42
# 4	65 - 3	42 + 23	65 – 23	65 - 43
# 5	65 – 4	41 + 24	65 – 24	65 - 44
# 6	65 – 5	40 + 25	65 – 25	65 - 45
# 7	65 – 6	39 + 26	65 – 26	65 - 46
# 8	65 – 7	38 + 27	65 – 27	65 - 47
# 9	65 – 8	37 + 28	65 – 28	65 - 48
# 10	65 – 9	36 + 29	65 – 29	65 - 49
# 11	65 – 10	35 + 30	65 – 30	65 – 50
# 12	65 – 11	34 + 31	65 – 31	65 - 51
# 13	65 – 12	33 + 32	65 – 32	65 – 52
# 14	65 – 13	32 + 33	65 – 33	65 - 53
# 15	65 – 14	31 + 34	65 – 34	65 – 54
# 16	65 – 15	30 + 35	65 – 35	65 - 55
# 17	65 – 16	29 + 36	65 – 36	65 – 56
# 18	65 – 17	28 + 37	65 – 37	65 – 57
# 19	65 – 18	27 + 38	65 – 38	65 - 58
# 20	65 – 19	26 + 39	65 – 39	65 – 59

Summary 3 Chapter 1 Lesson 9
Number 66

37 + 29 =30 + 7 + 20 + 9 = 30 + 20 + 7 + 9 = 50 + 16 = 66
66 – 29 = 66 – 20 – 9 = 46 – 9 = 37
27 + 39 = 20 + 7 + 30 + 9 = 20 + 30 + 7 + 9 = 50 + 16 = 66
66 – 39 = 66 – 30 – 9 = 36 – 9 = 27

Add or Subtract Numbers in row

	a	b	c	d
# 1	66 – 0	46 + 20	66 – 20	66 - 40
# 2	66 – 1	45 + 21	66 – 21	66 - 41
# 3	66 – 2	44 + 22	66 – 22	66 - 42
# 4	66 – 3	43 + 23	66 – 23	66 - 43
# 5	66 – 4	42 + 24	66 – 24	66 - 44
# 6	66 – 5	41 + 25	66 – 25	66 - 45
# 7	66 – 6	40 + 26	66 – 26	66 - 46
# 8	66 – 7	39 + 27	66 – 27	66 - 47
# 9	66 – 8	38 + 28	66 – 28	66 - 48
# 10	66 – 9	37 + 29	66 – 29	66 - 49
# 11	66 – 10	36 + 30	66 – 30	66 – 50
# 12	66 – 11	35 + 31	66 – 31	66 - 51
# 13	66 – 12	34 + 32	66 – 32	66 – 52
# 14	66 – 13	33 + 33	66 – 33	66 - 53
# 15	66 – 14	32 + 34	66 – 34	66 – 54
# 16	66 – 15	31 + 35	66 – 35	66 - 55
# 17	66 – 16	30 + 36	66 – 36	66 – 56
# 18	66 – 17	29 + 37	66 – 37	66 – 57
# 19	66 – 18	28 + 38	66 – 38	66 - 58
# 20	66 – 19	27 + 39	66 – 39	66 – 59

Summary 3 Chapter 1 Lesson 9
Number 67

38 + 29 =30 + 8 + 20 + 9 = 30 + 20 + 8 + 9 = 50 + 17 = 67

67 – 29 = 67 – 20 – 9 = 47 – 9 = 38

28 + 39 = 20 + 8 + 30 + 9 = 20 + 30 + 8 + 9 = 50 + 17 = 67

67 – 39 = 67 – 30 – 9 = 37 – 9 = 28

Add or Subtract Numbers in row

	a	b	c	d
# 1	67 – 0	47 + 20	67 – 20	67 - 40
# 2	67 – 1	46 + 21	67 – 21	67 - 41
# 3	67 – 2	45 + 22	67 – 22	67 - 42
# 4	67 – 3	44 + 23	67 – 23	67 - 43
# 5	67 – 4	43 + 24	67 – 24	67 - 44
# 6	67 – 5	42 + 25	67 – 25	67 - 45
# 7	67 – 6	41 + 26	67 – 26	67 - 46
# 8	67 – 7	40 + 27	67 – 27	67 - 47
# 9	67 – 8	39 + 28	67 – 28	67 - 48
# 10	67 – 9	38 + 29	67 – 29	67 - 49
# 11	67 – 10	37 + 30	67 – 30	67 – 50
# 12	67 – 11	36 + 31	67 – 31	67 - 51
# 13	67 – 12	35 + 32	67 – 32	67 – 52
# 14	67 – 13	34 + 33	67 – 33	67 - 53
# 15	67 – 14	33 + 34	67 – 34	67 – 54
# 16	67 – 15	32 + 35	67 – 35	67 - 55
# 17	67 – 16	31 + 36	67 – 36	67 – 56
# 18	67 – 17	30 + 37	67 – 37	67 – 57
# 19	67 – 18	29 + 38	67 – 38	67 - 58
# 20	67 – 19	28 + 39	67 – 39	67 – 59

Summary 3 Chapter 1 Lesson 10

Number 68

39 + 29 =30 + 9 + 20 + 9 = 30 + 20 + 9 + 9 = 50 + 18 = 68

68 – 29 = 68 – 20 – 9 = 48 – 9 = 39

29 + 39 = 20 + 9 + 30 + 9 = 20 + 30 + 9 + 9 = 50 + 18 = 68

68 – 39 = 68 – 30 – 9 = 38 – 9 = 29

Add or Subtract Numbers in row

	a	b	c	d
# 1	68 – 0	48 + 20	68 – 20	68 - 40
# 2	68 – 1	47 + 21	68 – 21	68 - 41
# 3	68 – 2	46 + 22	68 – 22	68 - 42
# 4	68 – 3	45 + 23	68 – 23	68 - 43
# 5	68 – 4	44 + 24	68 – 24	68 - 44
# 6	68 – 5	43 + 25	68 – 25	68 - 45
# 7	68 – 6	42 + 26	68 – 26	68 - 46
# 8	68 – 7	41 + 27	68 – 27	68 - 47
# 9	68 – 8	40 + 28	68 – 28	68 - 48
# 10	68 – 9	39 + 29	68 – 29	68 - 49
# 11	68 – 10	38 + 30	68 – 30	68 – 50
# 12	68 – 11	37 + 31	68 – 31	68 - 51
# 13	68 – 12	36 + 32	68 – 32	68 – 52
# 14	68 – 13	35 + 33	68 – 33	68 - 53
# 15	68 – 14	34 + 34	68 – 34	68 – 54
# 16	68 – 15	33 + 35	68 – 35	68 - 55
# 17	68 – 16	32 + 36	68 – 36	68 – 56
# 18	68 – 17	31 + 37	68 – 37	68 – 57
# 19	68 – 18	30 + 38	68 – 38	68 - 58
# 20	68 – 19	29 + 39	68 – 39	68 – 59

Summary 3 Chapter 1 Lesson 10
Number 69

40 + 29 =40 + 20 + 9 = 60 + 9 = 69
69 – 29 = 69 – 20 – 9 = 49 – 9 = 40
30 + 39 = 30 + 30 + 9 = 60 + 9 = 69
69 – 39 = 69 – 30 – 9 = 39 – 9 = 30

Add or Subtract Numbers in row

	a	b	c	d
# 1	69 – 0	49 + 20	69 – 20	69 - 40
# 2	69 – 1	48 + 21	69 – 21	69 - 41
# 3	69 – 2	47 + 22	69 – 22	69 - 42
# 4	69 – 3	46 + 23	69 – 23	69 - 43
# 5	69 – 4	45 + 24	69 – 24	69 - 44
# 6	69 – 5	44 + 25	69 – 25	69 - 45
# 7	69 – 6	43 + 26	69 – 26	69 - 46
# 8	69 – 7	42 + 27	69 – 27	69 - 47
# 9	69 – 8	41 + 28	69 – 28	69 - 48
# 10	69 – 9	40 + 29	69 – 29	69 - 49
# 11	69 – 10	39 + 30	69 – 30	69 – 50
# 12	69 – 11	38 + 31	69 – 31	69 - 51
# 13	69 – 12	37 + 32	69 – 32	69 – 52
# 14	69 – 13	36 + 33	69 – 33	69 - 53
# 15	69 – 14	35 + 34	69 – 34	69 – 54
# 16	69 – 15	34 + 35	69 – 35	69 - 55
# 17	69 – 16	33 + 36	69 – 36	69 – 56
# 18	69 – 17	32 + 37	69 – 37	69 – 57
# 19	69 – 18	31 + 38	69 – 38	69 - 58
# 20	69 – 19	30 + 39	69 – 39	69 – 59

Summary 3 Chapter 1 Lesson 11
Test 1A
Add or Subtract Numbers in row

1) 40 + 29
2) 69 – 29
3) 68 – 29
4) 39 + 29
5) 38 + 29
6) 67 – 29
7) 66 – 29
8) 37 + 29
9) 36 + 29
10) 65 – 29

11) 64 – 29
12) 35 + 29
13) 34 + 29
14) 63 – 29
15) 62 – 29
16) 33 + 29
17) 32 + 29
18) 61 – 29
19) 60 – 29
20) 31 + 29

Summary 3 Chapter 1 Lesson 11
Test 1B
Add or Subtract Numbers in row

1) 30 + 39
2) 69 – 39
3) 68 – 39
4) 29 + 39
5) 28 + 39
6) 67 – 39
7) 66 – 39
8) 27 + 39
9) 26 + 39
10) 65 – 39

11) 64 – 39
12) 25 + 39
13) 24 + 39
14) 63 – 39
15) 62 – 39
16) 23 + 39
17) 22 + 39
18) 61 – 39
19) 60 – 38
20) 22 + 38

Summary 3 Answers Chapter 1

Pretest 3 1) 243,815 2) 908 3) 4004 4) 180,180 5) 24

Lesson 6 Number 60

	a	b	c	d		a	b	c	d
# 1	60,	60,	40,	20	# 11	50,	60,	30,	10
# 2	59,	60,	39,	19	# 12	49,	60,	29,	9
# 3	58,	60,	38,	18	# 13	48,	60,	28,	8
# 4	57,	60,	37,	17	# 14	47,	60,	27,	7
# 5	56,	60,	36,	16	# 15	46,	60,	26,	6
# 6	55,	60,	35,	15	# 16	45,	60,	25,	5
# 7	54,	60,	34,	14	# 17	44,	60,	24,	4
# 8	53,	60,	33,	13	# 18	43,	60,	23,	3
# 9	52,	60,	32,	12	# 19	42,	60,	22,	2
# 10	51,	60,	31,	11	# 20	41,	60,	21,	1

Lesson 6 Number 61

	a	b	c	d		a	b	c	d
# 1	61,	61,	41,	21	# 11	51,	61,	31,	11
# 2	60,	61,	40,	20	# 12	50,	61,	30,	10
# 3	59,	61,	39,	19	# 13	49,	61,	29,	9
# 4	58,	61,	38,	18	# 14	48,	61,	28,	8
# 5	57,	61,	37,	17	# 15	47,	61,	27,	7
# 6	56,	61,	36,	16	# 16	46,	61,	26,	6
# 7	55,	61,	35,	15	# 17	45,	61,	25,	5
# 8	54,	61,	34,	14	# 18	44,	61,	24,	4
# 9	53,	61,	33,	13	# 19	43,	61,	23,	3
# 10	52,	61,	32,	12	# 20	42,	61,	22,	2

Lesson 7 Number 62

	a	b	c	d		a	b	c	d
# 1	62,	62,	42,	22	# 11	52,	62,	32,	12
# 2	61,	62,	41,	21	# 12	51,	62,	31,	11
# 3	60,	62,	40,	20	# 13	50,	62,	30,	10
# 4	59,	62,	39,	19	# 14	49,	62,	29,	9
# 5	58,	62,	38,	18	# 15	48,	62,	28,	8
# 6	57,	62,	37,	17	# 16	47,	62,	27,	7
# 7	56,	62,	36,	16	# 17	46,	62,	26,	6
# 8	55,	62,	35,	15	# 18	45,	62,	25,	5
# 9	54,	62,	34,	14	# 19	44,	62,	24,	4
# 10	53,	62,	33,	13	# 20	43,	62,	23,	3

Summary 3 Answers Chapter 1

Lesson 7 Number 63

	a, b, c, d		a, b, c, d
# 1	63, 63, 43, 23	# 11	53, 63, 33, 13
# 2	62, 63, 42, 22	# 12	52, 63, 32, 12
# 3	61, 63, 41, 21	# 13	51, 63, 31, 11
# 4	60, 63, 40, 20	# 14	50, 63, 30, 10
# 5	59, 63, 39, 19	# 15	49, 63, 29, 9
# 6	58, 63, 38, 18	# 16	48, 63, 28, 8
# 7	57, 63, 37, 17	# 17	47, 63, 27, 7
# 8	56, 63, 36, 16	# 18	46, 63, 26, 6
# 9	55, 63, 35, 15	# 19	45, 63, 25, 5
# 10	54, 63, 34, 14	# 20	44, 63, 24, 4

Lesson 8 Number 64

	a, b, c, d		a, b, c, d
# 1	64, 64, 44, 24	# 11	54, 64, 34, 14
# 2	63, 64, 43, 23	# 12	53, 64, 33, 13
# 3	62, 64, 42, 22	# 13	52, 64, 32, 12
# 4	61, 64, 41, 21	# 14	51, 64, 31, 11
# 5	60, 64, 40, 20	# 15	50, 64, 30, 10
# 6	59, 64, 39, 19	# 16	49, 64, 29, 9
# 7	58, 64, 38, 18	# 17	48, 64, 28, 8
# 8	57, 64, 37, 17	# 18	47, 64, 27, 7
# 9	56, 64, 36, 16	# 19	46, 64, 26, 6
# 10	55, 64, 35, 15	# 20	45, 64, 25, 5

Lesson 8 Number 65

	a, b, c, d		a, b, c, d
# 1	65, 65, 45, 25	# 11	55, 65, 35, 15
# 2	64, 65, 44, 24	# 12	54, 65, 34, 14
# 3	63, 65, 43, 23	# 13	53, 65, 33, 13
# 4	62, 65, 42, 22	# 14	52, 65, 32, 12
# 5	61, 65, 41, 21	# 15	51, 65, 31, 11
# 6	60, 65, 40, 20	# 16	50, 65, 30, 10
# 7	59, 65, 39, 19	# 17	49, 65, 29, 9
# 8	58, 65, 38, 18	# 18	48, 65, 28, 8
# 9	57, 65, 37, 17	# 19	47, 65, 27, 7
# 10	56, 65, 36, 16	# 20	46, 65, 26, 6

Summary 3 Answers Chapter 1

Lesson 9 Number 66

	a, b, c, d		a, b, c, d
# 1	66, 66, 46, 26	# 11	56, 66, 36, 16
# 2	65, 66, 45, 25	# 12	55, 66, 35, 15
# 3	64, 66, 44, 24	# 13	54, 66, 34, 14
# 4	63, 66, 43, 23	# 14	53, 66, 33, 13
# 5	62, 66, 42, 22	# 15	52, 66, 32, 12
# 6	61, 66, 41, 21	# 16	51, 66, 31, 11
# 7	60, 66, 40, 20	# 17	50, 66, 30, 10
# 8	59, 66, 39, 19	# 18	49, 66, 29, 9
# 9	58, 66, 38, 18	# 19	48, 66, 28, 8
# 10	57, 66, 37, 17	# 20	47, 66, 27, 7

Lesson 9 Number 67

	a, b, c, d		a, b, c, d
# 1	67, 67, 47, 27	# 11	57, 67, 37, 17
# 2	66, 67, 46, 26	# 12	56, 67, 36, 16
# 3	65, 67, 45, 25	# 13	55, 67, 35, 15
# 4	64, 67, 44, 24	# 14	54, 67, 34, 14
# 5	63, 67, 43, 23	# 15	53, 67, 33, 13
# 6	62, 67, 42, 22	# 16	52, 67, 32, 12
# 7	61, 67, 41, 21	# 17	51, 67, 31, 11
# 8	60, 67, 40, 20	# 18	50, 67, 30, 10
# 9	59, 67, 39, 19	# 19	49, 67, 29, 9
# 10	58, 67, 38, 18	# 20	48, 67, 28, 8

Lesson 10 Number 68

	a, b, c, d		a, b, c, d
# 1	68, 68, 48, 28	# 11	58, 68, 38, 18
# 2	67, 68, 47, 27	# 12	57, 68, 37, 17
# 3	66, 68, 46, 26	# 13	56, 68, 36, 16
# 4	65, 68, 45, 25	# 14	55, 68, 35, 15
# 5	64, 68, 44, 24	# 15	54, 68, 34, 14
# 6	63, 68, 43, 23	# 16	53, 68, 33, 13
# 7	62, 68, 42, 22	# 17	52, 68, 32, 12
# 8	61, 68, 41, 21	# 18	51, 68, 31, 11
# 9	60, 68, 40, 20	# 19	50, 68, 30, 10
# 10	59, 68, 39, 19	# 20	49, 68, 29, 9

Summary 3 Answers Chapter 1

Lesson 10 Number 69

	a b c d		a b c d
# 1	69, 69, 49, 29	# 11	59, 69, 39, 19
# 2	68, 69, 48, 28	# 12	58, 69, 38, 18
# 3	67, 69, 47, 27	# 13	57, 69, 37, 17
# 4	66, 69, 46, 26	# 14	56, 69, 36, 16
# 5	65, 69, 45, 25	# 15	55, 69, 35, 15
# 6	64, 69, 44, 24	# 16	54, 69, 34, 14
# 7	63, 69, 43, 23	# 17	53, 69, 33, 13
# 8	62, 69, 42, 22	# 18	52, 69, 32, 12
# 9	61, 69, 41, 21	# 19	51, 69, 31, 11
# 10	60, 69, 40, 20	# 20	50, 69, 30, 10

Lesson 11 Test 1A Add or Subtract Numbers in Row

1)	69	11)	35
2)	40	12)	64
3)	39	13)	63
4)	68	14)	34
5)	67	15)	33
6)	38	16)	62
7)	37	17)	61
8)	66	18)	32
9)	65	19)	31
10)	36	20)	60

Lesson 11 Test 1B Add or Subtract Numbers in Row

1)	69	11)	25
2)	30	12)	64
3)	29	13)	63
4)	68	14)	24
5)	67	15)	23
6)	28	16)	62
7)	27	17)	61
8)	66	18)	22
9)	65	19)	22
10)	26	20)	60

Summary 3

Chapter 2 Addition or Subtraction Numbers (90-99)

Summary 3 Chapter 2 Contents
Numbers 90 – 99
Lesson 1 - 11
Basic Problems

Work with lesson 1, read rules and examples, 2) Solve the problems from contents, 3) Check answers in the end of chapter 1, 4) Work with the next lessons 2, 3, 4, 5, ..., 5) Solve Test 1 A, 6) Grading Scale: A =10 points, B=9points, C=8 points, D=7 points, F=0-6 points, 7) If you solve 10 problems correctly, you get an A, 8) If have an A, you can move on to next chapter, 9) If your grade is lower than an A, work with the mistakes, then solve Test 1B, 10) When you finish all of the chapters, solve Final Exam in last chapter. Lesson = L

L 1 Number 90 Solve the problems # 5 (a, b, c, d, e), # 10 (a, b, c, d, e)
L2 Number 91 Solve the problems # 5 (a, b, c, d, e), # 10 (a, b, c, d, e)

L3 Number 92 Solve the problems # 5 (a, b, c, d, e), # 10 (a, b, c, d, e)
L4 Number 93 Solve the problems # 5 (a, b, c, d, e), # 10 (a, b, c, d, e)

L5 Number 94 Solve the problems # 5 (a, b, c, d, e), # 10 (a, b, c, d, e)
L6 Number 95 Solve the problems # 5 (a, b, c, d, e), # 10 (a, b, c, d, e)

L7 Number 96 Solve the problems # 5 (a, b, c, d, e), # 10 (a, b, c, d, e)
L8 Numbers 97 Solve the problems # 5 (a, b, c, d, e), # 10 (a, b, c, d, e)

L9 Numbers 98 Solve the problems # 5 (a, b, c, d, e), # 10 (a, b, c, d, e)
L10 Numbers 99 Solve the problems # 5 (a, b, c, d, e), # 10 (a, b, c, d, e)

L11 Test 2 A, Test 2 B for 50 minutes (10 problems)
Solve Test 2 A # 6 (a, b), # 7 (a, b), # 8 (a, b), # 9 (a, b), # 10 (a, b)
Solve Test 2 B # 6 (a, b), # 7 (a, b), # 8 (a, b), # 9 (a, b), # 10 (a, b)

Answers Chapter 2

Summary 3 Chapter 2 Lesson 1
Number 90
9 tens – 1 ten = 8 tens   1 ten = 10 ones

1	8 10
56	90
+34	- 34
90	56

Add or Subtract Numbers in column

	a	b	c	d	e
# 1	80 +10	90 -10	60 + 30	90 - 30	90 - 50
# 2	79 +11	90 - 11	59 +31	90 - 31	90 - 51
# 3	78 +12	90 -12	58 + 32	90 - 32	90 - 52
# 4	77 +13	90 -13	57 + 33	90 - 33	90 - 53
# 5	76 +14	90 - 14	56 + 34	90 - 34	90 - 54
# 6	75 +15	90 - 15	55 + 35	90 - 35	90 - 55
# 7	74 + 16	90 - 16	54 + 36	90 - 36	90 - 56
# 8	73 + 17	90 - 17	53 + 37	90 - 37	90 - 57
# 9	72 + 18	90 - 18	52 + 38	90 - 38	90 - 58
# 10	71 + 19	90 - 19	51 + 39	90 - 39	90 - 59

Summary 3 Chapter 2 Lesson 2
Number 91
9 tens – 1 ten = 8 tens 1 ten = 10 ones

1	8 11
57	91
+34	- 34
91	57

Add or Subtract Numbers in column

	a	b	c	d	e
# 1	81 +10	91 -10	61 + 30	91 - 30	91 - 50
# 2	80 +11	91 - 11	60 +31	91 - 31	91 - 51
# 3	79 +12	91 -12	59 + 32	91 - 32	91 - 52
# 4	78 +13	91 -13	58 + 33	91 - 33	91 - 53
# 5	77 +14	91 - 14	57 + 34	91 - 34	91 - 54
# 6	76 +15	91 - 15	56 + 35	91 - 35	91 - 55
# 7	75 + 16	91 - 16	55 + 36	91 - 36	91 - 56
# 8	74 + 17	91 - 17	54 + 37	91 - 37	91 - 57
# 9	73 + 18	91 - 18	53 + 38	91 - 38	91 - 58
# 10	72 + 19	91 - 19	52 + 39	91 - 39	91 - 59

Summary 3 Chapter 2 Lesson 3
Number 92
9 tens – 1 ten = 8 tens   1 ten = 10 ones

1	8 12
58	92
+34	- 34
92	58

Add or Subtract Numbers in column

	a	b	c	d	e
# 1	82 +10	92 -10	62 + 30	92 - 30	92 - 50
# 2	81 +11	92 - 11	61 +31	92 - 31	92 - 51
# 3	80 +12	92 -12	60 + 32	92 - 32	92 - 52
# 4	79 +13	92 -13	59 + 33	92 - 33	92 - 53
# 5	78 +14	92 - 14	58 + 34	92 - 34	92 - 54
# 6	77 +15	92 - 15	57 + 35	92 - 35	92 - 55
# 7	76 + 16	92 - 16	56 + 36	92 - 36	92 - 56
# 8	75 + 17	92 - 17	55 + 37	92 - 37	92 - 57
# 9	74 + 18	92 - 18	54 + 38	92 - 38	92 - 58
# 10	73 + 19	92 - 19	53 + 39	92 - 39	92 - 59

Summary 3 Chapter 2 Lesson 4
Number 93

9 tens – 1 ten = 8 tens   1 ten = 10 ones

1	8 13
59	93
+34	- 34
93	59

Add or Subtract Numbers in column

	a	b	c	d	e
# 1	83 +10	93 -10	63 + 30	93 - 30	93 - 50
# 2	82 +11	93 - 11	62 +31	93 - 31	93 - 51
# 3	81 +12	93 -12	61 + 32	93 - 32	93 - 52
# 4	80 +13	93 -13	60 + 33	93 - 33	93 - 53
# 5	79 +14	93 - 14	59 + 34	93 - 34	93 - 54
# 6	78 +15	93 - 15	58 + 35	93 - 35	93 - 55
# 7	77 + 16	93 - 16	57 + 36	93 - 36	93 - 56
# 8	76 + 17	93 - 17	56 + 37	93 - 37	93 - 57
# 9	75 + 18	93 - 18	55 + 38	93 - 38	93 - 58
# 10	74 + 19	93 - 19	54 + 39	93 - 39	93 - 59

Summary 3 Chapter 2 Lesson 5

Number 94

9 tens – 1 ten = 8 tens 1 ten = 10 ones

1	8 14
55	94
+39	- 39
94	55

Add or Subtract Numbers in column

	a	b	c	d	e
# 1	84 +10	94 -10	64 + 30	94 - 30	94 - 50
# 2	83 +11	94 - 11	63 +31	94 - 31	94 - 51
# 3	82 +12	94 -12	62 + 32	94 - 32	94 - 52
# 4	81 +13	94 -13	61 + 33	94 - 33	94 - 53
# 5	80 +14	94 - 14	60 + 34	94 - 34	94 - 54
# 6	79 +15	94 - 15	59 + 35	94 - 35	94 - 55
# 7	78 + 16	94 - 16	58 + 36	94 - 36	94 - 56
# 8	77 + 17	94 - 17	57 + 37	94 - 37	94 - 57
# 9	76 + 18	94 - 18	56 + 38	94 - 38	94 - 58
# 10	75 + 19	94 - 19	55 + 39	94 - 39	94 - 59

Summary 3 Chapter 2 Lesson 6
Number 95

9 tens – 1 ten = 8 tens   1 ten = 10 ones

1	8 15
56	95
+39	- 39
95	56

Add or Subtract Numbers in column

	a	b	c	d	e
# 1	85 +10	95 -10	65 + 30	95 - 30	95 - 50
# 2	84 +11	95 - 11	64 +31	95 - 31	95 - 51
# 3	83 +12	95 -12	63 + 32	95 - 32	95 - 52
# 4	82 +13	95 -13	62 + 33	95 - 33	95 - 53
# 5	81 +14	95 - 14	61 + 34	95 - 34	95 - 54
# 6	80 +15	95 - 15	60 + 35	95 - 35	95 - 55
# 7	79 + 16	95 - 16	59 + 36	95 - 36	95 - 56
# 8	78 + 17	95 - 17	58 + 37	95 - 37	95 - 57
# 9	77 + 18	95 - 18	57 + 38	95 - 38	95 - 58
# 10	76 + 19	95 - 19	56 + 39	95 - 39	95 - 59

Summary 3 Chapter 2 Lesson 7
Number 96
9 tens – 1 ten = 8 tens   1 ten = 10 ones

1	8 16
57	96
+39	- 39
96	57

Add or Subtract Numbers in column

	a	b	c	d	e
# 1	86 +10	96 -10	66 + 30	96 - 30	96 - 50
# 2	85 +11	96 - 11	65 +31	96 - 31	96 - 51
# 3	84 +12	96 -12	64 + 32	96 - 32	96 - 52
# 4	83 +13	96 -13	63 + 33	96 - 33	96 - 53
# 5	82 +14	96 - 14	62 + 34	96 - 34	96 - 54
# 6	81 +15	96 - 15	61 + 35	96 - 35	96 - 55
# 7	80 + 16	96 - 16	60 + 36	96 - 36	96 - 56
# 8	79 + 17	96 - 17	59 + 37	96 - 37	96 - 57
# 9	78 + 18	96 - 18	58 + 38	96 - 38	96 - 58
# 10	77 + 19	96 - 19	57 + 39	96 - 39	96 - 59

Summary 3 Chapter 2 Lesson 8

Number 97

9 tens – 1 ten = 8 tens   1 ten = 10 ones

1	817
48	97
+49	- 49
97	48

Add or Subtract Numbers in Column

	a	b	c	d	e
# 1	87 +10	97 -10	67 + 30	97 - 30	97 - 50
# 2	86 +11	97 - 11	66 +31	97 - 31	97 - 51
# 3	85 +12	97 -12	65 + 32	97 - 32	97 - 52
# 4	84 +13	97 -13	64 + 33	97 - 33	97 - 53
# 5	83 +14	97 - 14	63 + 34	97 - 34	97 - 54
# 6	82 +15	97 - 15	62 + 35	97 - 35	97 - 55
# 7	81 + 16	97 - 16	61 + 36	97 - 36	97 - 56
# 8	80 + 17	97 - 17	60 + 37	97 - 37	97 - 57
# 9	79 + 18	97 - 18	59 + 38	97 - 38	97 - 58
# 10	78 + 19	97 - 19	58 + 39	97 - 39	97 - 59

Summary 3 Chapter 2 Lesson 9
Number 98
9 tens – 1 ten = 8 tens   1 ten = 10 ones

1	8 18
59	98
+39	- 39
98	59

Add or Subtract Numbers in column

	a	b	c	d	e
# 1	88 +10	98 -10	68 + 30	98 - 30	98 - 50
# 2	87 +11	98 - 11	67 +31	98 - 31	98 - 51
# 3	86 +12	98 -12	66 + 32	98 - 32	98 - 52
# 4	85 +13	98 -13	65 + 33	98 - 33	98 - 53
# 5	84 +14	98 - 14	64 + 34	98 - 34	98 - 54
# 6	83 +15	98 - 15	63 + 35	98 - 35	98 - 55
# 7	82 + 16	98 - 16	62 + 36	98 - 36	98 - 56
# 8	81 + 17	98 - 17	61 + 37	98 - 37	98 - 57
# 9	80 + 18	98 - 18	60 + 38	98 - 38	98 - 58
# 10	79 + 19	98 - 19	59 + 39	98 - 39	98 - 59

Summary 3 Chapter 2 Lesson 10
Number 99

65	99
+34	- 34
99	65

Add or Subtract Numbers in column

	a	b	c	d	e
# 1	89 +10	99 -10	69 + 30	99 - 30	99 - 50
# 2	88 +11	99 - 11	68 +31	99 - 31	99 - 51
# 3	87 +12	99 -12	67 + 32	99 - 32	99 - 52
# 4	86 +13	99 -13	66 + 33	99 - 33	99 - 53
# 5	85 +14	99 - 14	65 + 34	99 - 34	99 - 54
# 6	84 +15	99 - 15	64 + 35	99 - 35	99 - 55
# 7	83 + 16	99 - 16	63 + 36	99 - 36	99 - 56
# 8	82 + 17	99 - 17	62 + 37	99 - 37	99 - 57
# 9	81 + 18	99 - 18	61 + 38	99 - 38	99 - 58
# 10	80 + 19	99 - 19	60 + 39	99 - 39	99 - 59

Summary 3 Chapter 2 Lesson 11
Test 2 A
Add or Subtract Numbers in Column

1) a) 56 +34 b) 90 - 34

2) a) 46 +44 b) 90 - 44

3) a) 57 +34 b) 91 - 34

4) a) 47 +44 b) 91 - 44

5) a) 58 +34 b) 92 - 34

6) a) 48 +44 b) 92 - 44

7) a) 59 +34 b) 93 - 34

8) a) 49 +44 b) 93 - 44

9) a) 55 +39 b) 94 - 39

10) a) 45 +49 b) 94 - 49

Summary 3 Chapter 2 Lesson 11
Test 2 B
Add or Subtract Numbers in Column

1) a) 56 + 39   b) 95 - 39

2) a) 57 + 38   b) 95 - 38

3) a) 57 + 39   b) 96 - 39

4) a) 47 + 49   b) 96 - 49

5) a) 48 + 49   b) 97 - 49

6) a) 58 + 39   b) 97 - 39

7) a) 59 + 39   b) 98 - 39

8) a) 49 + 49   b) 98 - 49

9) a) 65 + 34   b) 99 - 34

10) a) 60 + 39   b) 99 - 39

Summary 3 Answers Chapter 2

Lesson 1 Number 90

	a, b, c, d, e
# 1	90, 80, 90, 60, 40
# 2	90, 79, 90, 59, 39
# 3	90, 78, 90, 58, 38
# 4	90, 77, 90, 57, 37
# 5	90, 76, 90, 56, 36
# 6	90, 75, 90, 55, 35
# 7	90, 74, 90, 54, 34
# 8	90, 73, 90, 53, 33
# 9	90, 72, 90, 52, 32
# 10	90, 71, 90, 51, 31

Lesson 2 Number 91

	a, b, c, d, e
# 1	91, 81, 91, 61, 41
# 2	91, 80, 91, 60, 40
# 3	91, 79, 91, 59, 39
# 4	91, 78, 91, 58, 38
# 5	91, 77, 91, 57, 37
# 6	91, 76, 91, 56, 36
# 7	91, 75, 91, 55, 35
# 8	91, 74, 91, 54, 34
# 9	91, 73, 91, 53, 33
# 10	91, 72, 91, 52, 32

Lesson 3 Number 92

	a, b, c, d, e
# 1	92, 82, 92, 62, 42
# 2	92, 81, 92, 61, 41
# 3	92, 80, 92, 60, 40
# 4	92, 79, 92, 59, 39
# 5	92, 78, 92, 58, 38
# 6	92, 77, 92, 57, 37
# 7	92, 76, 92, 56, 36
# 8	92, 75, 92, 55, 35
# 9	92, 74, 92, 54, 34
# 10	92, 73, 92, 53, 33

Lesson 4 Number 93

	a, b, c, d, e
# 1	93, 83, 93, 63, 43
# 2	93, 82, 93, 62, 42
# 3	93, 81, 93, 61, 41
# 4	93, 80, 93, 60, 40
# 5	93, 79, 93, 59, 39
# 6	93, 78, 93, 58, 38
# 7	93, 77, 93, 57, 37
# 8	93, 76, 93, 56, 36
# 9	93, 75, 93, 55, 35
# 10	93, 74, 93, 54, 34

Lesson 5 Number 94

	a, b, c, d, e
# 1	94, 84, 94, 64, 44
# 2	94, 83, 94, 63, 43
# 3	94, 82, 94, 62, 42
# 4	94, 81, 94, 61, 41
# 5	94, 80, 94, 60, 40
# 6	94, 79, 94, 59, 39
# 7	94, 78, 94, 58, 38
# 8	94, 77, 94, 57, 37
# 9	94, 76, 94, 56, 36
# 10	94, 75, 94, 55, 35

Lesson 6 Number 95

	a, b, c, d, e
# 1	95, 85, 95, 65, 45
# 2	95, 84, 95, 64, 44
# 3	95, 83, 95, 63, 43
# 4	95, 82, 95, 62, 42
# 5	95, 81, 95, 61, 41
# 6	95, 80, 95, 60, 40
# 7	95, 79, 95, 59, 39
# 8	95, 78, 95, 58, 38
# 9	95, 77, 95, 57, 37
# 10	95, 76, 95, 56, 36

Summary 3 Answers Chapter 2

Lesson 7 Number 96

	a, b, c, d, e
# 1	96, 86, 96, 66, 46
# 2	96, 85, 96, 65, 45
# 3	96, 84, 96, 64, 44
# 4	96, 83, 96, 63, 43
# 5	96, 82, 96, 62, 42
# 6	96, 81, 96, 61, 41
# 7	96, 80, 96, 60, 40
# 8	96, 79, 96, 59, 39
# 9	96, 78, 96, 58, 38
# 10	96, 77, 96, 57, 37

Lesson 8 Number 97

	a, b, c, d, e
# 1	97, 87, 97, 67, 47
# 2	97, 86, 97, 66, 46
# 3	97, 85, 97, 65, 45
# 4	97, 84, 97, 64, 44
# 5	97, 83, 97, 63, 43
# 6	97, 82, 97, 62, 42
# 7	97, 81, 97, 61, 41
# 8	97, 80, 97, 60, 40
# 9	97, 79, 97, 59, 39
# 10	97, 78, 97, 58, 38

Lesson 9 Number 98

	a, b, c, d, e
# 1	98, 88, 98, 68, 48
# 2	98, 87, 98, 67, 47
# 3	98, 86, 98, 66, 46
# 4	98, 85, 98, 65, 45
# 5	98, 84, 98, 64, 44
# 6	98, 83, 98, 63, 43
# 7	98, 82, 98, 62, 42
# 8	98, 81, 98, 61, 41
# 9	98, 80, 98, 60, 40
# 10	98, 79, 98, 59, 39

Lesson 10 Number 99

	a, b, c, d, e
# 1	99, 89, 99, 69, 49
# 2	99, 88, 99, 68, 48
# 3	99, 87, 99, 67, 47
# 4	99, 86, 99, 66, 46
# 5	99, 85, 99, 65, 45
# 6	99, 84, 99, 64, 44
# 7	99, 83, 99, 63, 43
# 8	99, 82, 99, 62, 42
# 9	99, 81, 99, 61, 41
# 10	99, 80, 99, 60, 40

Lesson 11 Test 2 A

1)	a)	90	b)	56
2)	a)	90	b)	46
3)	a)	91	b)	57
4)	a)	91	b)	47
5)	a)	92	b)	58
6)	a)	92	b)	48
7)	a)	93	b)	59
8)	a)	93	b)	49
9)	a)	94	b)	55
10)	a)	94	b)	45

Lesson 11 Test 2 B

1)	a)	95	b)	56
2)	a)	95	b)	57
3)	a)	96	b)	57
4)	a)	96	b)	47
5)	a)	97	b)	48
6)	a)	97	b)	58
7)	a)	98	b)	59
8)	a)	98	b)	48
9)	a)	99	b)	65
10)	a)	99	b)	60

Summary 3

Chapter 3 Addition or Subtraction Numbers (70-79)

Summary 3 Chapter 3 Contents
Numbers 70 – 79
Lesson 1 - 13
Basic Problems

Work with lesson 1, read rules and examples, 2) Solve the problems from contents, 3) Check answers in the end of chapter 1, 4) Work with the next lessons 2, 3, 4, 5, ..., 5) Solve Test 1 A, 6) Grading Scale: A =10 points, B=9points, C=8 points, D=7 points, F=0-6 points, 7) If you solve 10 problems correctly, you get an A, 8) If have an A, you can move on to next chapter, 9) If your grade is lower than an A, work with the mistakes, then solve Test 1B, 10) When you finish all of the chapters, solve Final Exam in last chapter. Lesson = L

L1 Number 70 Solve the problems # 5 (a, b, c, d, e), # 10 (a, b, c, d, e)
L2 Number 71 Solve the problems # 5 (a, b, c, d, e), # 10 (a, b, c, d, e)
L3 Number 72 Solve the problems # 5 (a, b, c, d, e), # 10 (a, b, c, d, e)
L4 Number 73 Solve the problems # 5 (a, b, c, d, e), # 10 (a, b, c, d, e)

L5 Number 74 Solve the problems # 5 (a, b, c, d, e), # 10 (a, b, c, d, e)
L6 Number 75 Solve the problems # 5 (a, b, c, d, e), # 10 (a, b, c, d, e)
L7 Number 76 Solve the problems # 5 (a, b, c, d, e), # 10 (a, b, c, d, e)
L8 Numbers 77 Solve the problems # 5 (a, b, c, d, e), # 10 (a, b, c, d, e)

L9 Numbers 78 Solve the problems # 5 (a, b, c, d, e), # 10 (a, b, c, d, e)
L10 Numbers 79 Solve the problems # 5 (a, b, c, d, e), # 10 (a, b, c, d, e)
L11 Test 3 A, Test 3 B for 50 minutes
Solve Test 3 A # 6(a, b), # 7(a, b), # 8(a, b), # 9(a, b), # 10 (a, b)
Solve Test 3 B # 6(a, b), # 7(a, b), # 8(a, b), # 9(a, b), # 10 (a, b)

L12 Review for Final Exam 3
Review 3 A # 1 – 20 odd or #1, 3, 5, 7, 9, 11, 13, 15, 17, 19
Review 3 B # 1 – 20 odd or #1, 3, 5, 7, 9, 11, 13, 15, 17, 19
L13 Final Exam 3
Final Exam 3 A # 1 – 20 odd or #1, 3, 5, 7, 9, 11, 13, 15, 17, 19
Final Exam 3 B # 1 – 20 odd or #1, 3, 5, 7, 9, 11, 13, 15, 17, 19
Grade 3 Answers Chapter 3

Summary 3 Chapter 3 Lesson 1

Number 70

7 tens – 1 ten = 6 tens   1 ten = 10 ones

1	6 10
36	70
+34	- 34
70	36

Add or Subtract Numbers in column

	a	b	c	d	e
# 1	60 +10	70 -10	40 + 30	70 - 30	70 - 50
# 2	59 +11	70 - 11	39 +31	70 - 31	70 - 51
# 3	58 +12	70 -12	38 + 32	70 - 32	70 - 52
# 4	57 +13	70 -13	37 + 33	70 - 33	70 - 53
# 5	56 +14	70 - 14	36 + 34	70 - 34	70 - 54
# 6	55 +15	70 - 15	35 + 35	70 - 35	70 - 55
# 7	54 + 16	70 - 16	34 + 36	70 - 36	70 - 56
# 8	53 + 17	70 - 17	33 + 37	70 - 37	70 - 57
# 9	52 + 18	70 - 18	32 + 38	70 - 38	70 - 58
# 10	51 + 19	70 - 19	31 + 39	70 - 39	70 - 59

Summary 3 Chapter 3 Lesson 2

Number 71

7 tens – 1 ten = 6 tens 1 ten = 10 ones

1	6 11
37	71
+34	- 34
71	37

Add or Subtract Numbers in column

	a	b	c	d	e
# 1	61 +10	71 -10	41 + 30	71 - 30	71 - 50
# 2	60 +11	71 - 11	40 +31	71 - 31	71 - 51
# 3	59 +12	71 -12	39 + 32	71 - 32	71 - 52
# 4	58 +13	71 -13	38 + 33	71 - 33	71 - 53
# 5	57 +14	71 - 14	37 + 34	71 - 34	71 - 54
# 6	56 +15	71 - 15	36 + 35	71 - 35	71 - 55
# 7	55 + 16	71 - 16	35 + 36	71 - 36	71 - 56
# 8	54 + 17	71 - 17	34 + 37	71 - 37	71 - 57
# 9	53 + 18	71 - 18	33 + 38	71 - 38	71 - 58
# 10	52 + 19	71 - 19	32 + 39	71 - 39	71 - 59

Summary 3 Chapter 3 Lesson 3
Number 72
7 tens – 1 ten = 6 tens   1 ten = 10 ones

1	6 12
38	72
+34	- 34
72	38

Add or Subtract Numbers in column

	a	b	c	d	e
# 1	62 +10	72 -10	42 + 30	72 - 30	72 - 50
# 2	61 +11	72 - 11	41 +31	72 - 31	72 - 51
# 3	60 +12	72 -12	40 + 32	72 - 32	72 - 52
# 4	59 +13	72 -13	39 + 33	72 - 33	72 - 53
# 5	58 +14	72 - 14	38 + 34	72 - 34	72 - 54
# 6	57 +15	72 - 15	37 + 35	72 - 35	72 - 55
# 7	56 + 16	72 - 16	36 + 36	72 - 36	72 - 56
# 8	55 + 17	72 - 17	35 + 37	72 - 37	72 - 57
# 9	54 + 18	72 - 18	34 + 38	72 - 38	72 - 58
# 10	53 + 19	72 - 19	33 + 39	72 - 39	72 - 59

Summary 3 Chapter 3 Lesson 4

Number 73

7 tens – 1 ten = 6 tens   1 ten = 10 ones

```
   1        6 13
  39          73
 +34        - 34
  73          39
```

Add or Subtract Numbers in column

	a	b	c	d	e
# 1	63 +10	73 -10	43 + 30	73 - 30	73 - 50
# 2	62 +11	73 - 11	42 +31	73 - 31	73 - 51
# 3	61 +12	73 -12	41 + 32	73 - 32	73 - 52
# 4	60 +13	73 -13	40 + 33	73 - 33	73 - 53
# 5	59 +14	73 - 14	39 + 34	73 - 34	73 - 54
# 6	58 +15	73 - 15	38 + 35	73 - 35	73 - 55
# 7	57 + 16	73 - 16	37 + 36	73 - 36	73 - 56
# 8	56 + 17	73 - 17	36 + 37	73 - 37	73 - 57
# 9	55 + 18	73 - 18	35 + 38	73 - 38	73 - 58
# 10	54 + 19	73 - 19	34 + 39	73 - 39	73 - 59

Summary 3 Chapter 3 Lesson 5
Number 74

7 tens – 1 ten = 6 tens 1 ten = 10 ones

1	6 14
35	74
+39	- 39
74	35

Add or Subtract Numbers in column

	a	b	c	d	e
# 1	64 +10	74 -10	44 + 30	74 - 30	74 - 50
# 2	63 +11	74 - 11	43 +31	74 - 31	74 - 51
# 3	62 +12	74 -12	42 + 32	74 - 32	74 - 52
# 4	61 +13	74 -13	41 + 33	74 - 33	74 - 53
# 5	60 +14	74 - 14	40 + 34	74 - 34	74 - 54
# 6	59 +15	74 - 15	39 + 35	74 - 35	74 - 55
# 7	58 +16	74 - 16	38 + 36	74 - 36	74 - 56
# 8	57 + 17	74 - 17	37 + 37	74 - 37	74 - 57
# 9	56 + 18	74 - 18	36 + 38	74 - 38	74 - 58
# 10	55 + 19	74 - 19	35 + 39	74 - 39	74 - 59

Summary 3 Chapter 3 Lesson 6

Number 75

7 tens – 1 ten = 6 tens   1 ten = 10 ones

1	6 15
36	75
+39	- 39
75	36

Add or Subtract Numbers in column

	a	b	c	d	e
# 1	65 +10	75 -10	45 + 30	75 - 30	75 - 50
# 2	64 +11	75 - 11	44 +31	75 - 31	75 - 51
# 3	63 +12	75 -12	43 + 32	75 - 32	75 - 52
# 4	62 +13	75 -13	42 + 33	75 - 33	75 - 53
# 5	61 +14	75 - 14	41 + 34	75 - 34	75 - 54
# 6	60 +15	75 - 15	40 + 35	75 - 35	75 - 55
# 7	59 + 16	75 - 16	39 + 36	75 - 36	75 - 56
# 8	58 + 17	75 - 17	38 + 37	75 - 37	75 - 57
# 9	57 + 18	75 - 18	37 + 38	75 - 38	75 - 58
# 10	56 + 19	75 - 19	36 + 39	75 - 39	75 - 59

Summary 3 Chapter 3 Lesson 7

Number 76

7 tens – 1 ten = 6 tens   1 ten = 10 ones

1	6 16
37	76
+39	- 39
76	37

Add or Subtract Numbers in column

	a	b	c	d	e
# 1	66 +10	76 -10	46 + 30	76 - 30	76 - 50
# 2	65 +11	76 - 11	45 +31	76 - 31	76 - 51
# 3	64 +12	76 -12	44 + 32	76 - 32	76 - 52
# 4	63 +13	76 -13	43 + 33	76 - 33	76 - 53
# 5	62 +14	76 - 14	42 + 34	76 - 34	76 - 54
# 6	61 +15	76 - 15	41 + 35	76 - 35	76 - 55
# 7	60 + 16	76 - 16	40 + 36	76 - 36	76 - 56
# 8	59 + 17	76 - 17	39 + 37	76 - 37	76 - 57
# 9	58 + 18	76 - 18	38 + 38	76 - 38	76 - 58
# 10	57 + 19	76 - 19	37 + 39	76 - 39	76 - 59

Summary 3 Chapter 3 Lesson 8
Number 77
7 tens – 1 ten = 6 tens   1 ten = 10 ones

1	617
38	77
+39	- 39
77	38

Add or Subtract Numbers in Column

	a	b	c	d	e
# 1	67 +10	77 -10	47 + 30	77 - 30	77 - 50
# 2	66 +11	77 - 11	46 +31	77 - 31	77 - 51
# 3	65 +12	77 -12	45 + 32	77 - 32	77 - 52
# 4	64 +13	77 -13	44 + 33	77 - 33	77 - 53
# 5	63 +14	77 - 14	43 + 34	77 - 34	77 - 54
# 6	62 +15	77 - 15	42 + 35	77 - 35	77 - 55
# 7	61 + 16	77 - 16	41 + 36	77 - 36	77 - 56
# 8	60 + 17	77 - 17	40 + 37	77 - 37	77 - 57
# 9	59 + 18	77 - 18	39 + 38	77 - 38	77 - 58
# 10	58 + 19	77 - 19	38 + 39	77 - 39	77 - 59

Summary 3 Chapter 3 Lesson 9
Number 78
7 tens – 1 ten = 6 tens   1 ten = 10 ones

1	6 18
39	78
+39	- 39
78	39

Add or Subtract Numbers in column

	a	b	c	d	e
# 1	68 +10	78 -10	48 + 30	78 - 30	78 - 50
# 2	67 +11	78 - 11	47 +31	78 - 31	78 - 51
# 3	66 +12	78 -12	46 + 32	78 - 32	78 - 52
# 4	65 +13	78 -13	45 + 33	78 - 33	78 - 53
# 5	64 +14	78 - 14	44 + 34	78 - 34	78 - 54
# 6	63 +15	78 - 15	43 + 35	78 - 35	78 - 55
# 7	62 + 16	78 - 16	42 + 36	78 - 36	78 - 56
# 8	61 + 17	78 - 17	41 + 37	78 - 37	78 - 57
# 9	60 + 18	78 - 18	40 + 38	78 - 38	78 - 58
# 10	59 + 19	78 - 19	49 + 39	78 - 39	78 - 59

Summary 3 Chapter 3 Lesson 10
Number 79

7 tens – 1 ten = 6 tens 1 ten = 10 ones

45	79
+34	- 34
79	45

Add or Subtract Numbers in column

	a	b	c	d	e
# 1	69 +10	79 -10	49 + 30	79 - 30	79 - 50
# 2	68 +11	79 - 11	48 +31	79 - 31	79 - 51
# 3	67 +12	79 -12	47 + 32	79 - 32	79 - 52
# 4	66 +13	79 -13	46 + 33	79 - 33	79 - 53
# 5	65 +14	79 - 14	45 + 34	79 - 34	79 - 54
# 6	64 +15	79 - 15	44 + 35	79 - 35	79 - 55
# 7	63 + 16	79 - 16	43 + 36	79 - 36	79 - 56
# 8	62 + 17	79 - 17	42 + 37	79 - 37	79 - 57
# 9	61 + 18	79 - 18	41 + 38	79 - 38	79 - 58
# 10	60 + 19	79 - 19	40 + 39	79 - 39	79 - 59

Summary 3 Chapter 3 Lesson 11
Test 3 A
Add or Subtract Numbers in Column

1)	a)	36 +34	b)	70 - 35
2)	a)	26 +44	b)	70 - 34
3)	a)	37 +34	b)	71 - 34
4)	a)	27 +44	b)	71 - 33
5)	a)	38 +34	b)	72 - 33
6)	a)	28 +44	b)	72 - 47
7)	a)	39 +34	b)	73 - 47
8)	a)	29 +44	b)	73 - 46
9)	a)	35 +39	b)	74 - 46
10)	a)	25 +49	b)	74 - 45

Summary 3 Chapter 3 Lesson 11
Test 3 B
Add or Subtract Numbers in Column

	a)	b)
1)	36 +39	75 - 50
2)	37 +38	75 - 49
3)	37 +39	76 - 49
4)	27 +49	76 - 48
5)	28 +49	77 - 48
6)	38 +39	77 - 42
7)	39 +39	78 - 42
8)	29 +49	78 - 41
9)	45 +34	79 - 41
10)	40 +39	79 - 40

Summary 3 Chapter 3 Lesson 12
Review 3 A
Add or Subtract numbers in column

# 1	53 + 37	# 6	93 -58
# 2	69 +25	# 7	97 -39
# 3	69 +29	# 8	91 -67
# 4	56 +36	# 9	95 - 47
# 5	57 +39	# 10	99 - 59

Add or Subtract numbers in row

11) 73 + 18
12) 74 + 19

13) 66 + 29
14) 79 + 18

15) 69 + 21
16) 90 – 71

17) 92 – 64
18) 94 – 49

19) 96 – 57
20) 93 - 37

Summary 3 Chapter 3 Lesson 12
Review 3 B
Add or Subtract numbers in column

# 1	67 + 24	# 6	92 - 58
# 2	67 +28	# 7	96 - 39
# 3	68 +31	# 8	90 -65
# 4	59 +34	# 9	94 - 57
# 5	58 +39	# 10	98 - 29

Add or Subtract numbers in row

11) 29 + 61
12) 59 + 33

13) 49 + 45
14) 67 + 29

15) 66 + 29
16) 91 – 29

17) 93 – 34
18) 95 – 29

19) 97 – 48
20) 97 – 41

Summary 3 Chapter 3 Lesson 13

Final Exam 3 A

Add or Subtract Numbers in Row

# 1	31 + 59
# 2	69 + 22
# 3	13 + 79
# 4	69 + 24
# 5	35 + 59
# 6	90 – 31
# 7	91 – 69
# 8	92 – 13
# 9	93 – 69
# 10	94 – 35

Add or Subtract Numbers in Column

# 11 46	# 12 39	# 13 68	# 14 19	# 15 80
+49	+57	+ 29	+79	+ 19
# 16 95	# 17 96	# 18 97	# 19 98	# 20 99
-49	-57	-29	-79	- 19

Summary 3 Chapter 3 Lesson 13

Final Exam 3 B
1-20 odd or # 1, 3, 5, 7, 9, 11, 13, 15, 17, 19

Add or Subtract Numbers in Row

# 1	41 + 49
# 2	59 + 32
# 3	23 + 69
# 4	59 + 34
# 5	25 + 69
# 6	90 – 41
# 7	91 – 59
# 8	92 – 23
# 9	93 – 59
# 10	94 – 45

Add or Subtract Numbers in Column

# 11 36	# 12 49	# 13 58	# 14 29	# 15 70
+59	+47	+ 39	+69	+ 29
# 16 95	# 17 96	# 18 97	# 19 98	# 20 99
-59	-67	-39	-69	- 29

Summary 3 Answers Chapter 3

Lesson 1 Number 70

	a	b	c	d	e
# 1	70,	60,	70,	40,	20
# 2	70,	59,	70,	39,	19
# 3	70,	58,	70,	38,	18
# 4	70,	57,	70,	37,	17
# 5	70,	56,	70,	36,	16
# 6	70,	55,	70,	35,	15
# 7	70,	54,	70,	34,	14
# 8	70,	53,	70,	33,	13
# 9	70,	52,	70,	32,	12
# 10	70,	51,	70,	31,	11

Lesson 2 Number 71

	a	b	c	d	e
# 1	71,	61,	71,	41,	21
# 2	71,	60,	71,	40,	20
# 3	71,	59,	71,	39,	19
# 4	71,	58,	71,	38,	18
# 5	71,	57,	71,	37,	17
# 6	71,	56,	71,	36,	16
# 7	71,	55,	71,	35,	15
# 8	71,	54,	71,	34,	14
# 9	71,	53,	71,	33,	13
# 10	71,	52,	71,	32,	12

Lesson 3 Number 72

	a	b	c	d	e
# 1	72,	62,	72,	42,	22
# 2	72,	61,	72,	41,	21
# 3	72,	60,	72,	40,	20
# 4	72,	59,	72,	39,	19
# 5	72,	58,	72,	38,	18
# 6	72,	57,	72,	37,	17
# 7	72,	56,	72,	36,	16
# 8	72,	55,	72,	35,	15
# 9	72,	54,	72,	34,	14
# 10	72,	53,	72,	33,	13

Summary 3 Answers Chapter 3

Lesson 4 Number 73

	a	b	c	d	e
# 1	73,	63,	73,	43,	23
# 2	73,	62,	73,	42,	22
# 3	73,	61,	73,	41,	21
# 4	73,	60,	73,	40,	20
# 5	73,	59,	73,	39,	19
# 6	73,	58,	73,	38,	18
# 7	73,	57,	73,	37,	17
# 8	73,	56,	73,	36,	16
# 9	73,	55,	73,	35,	15
# 10	73,	54,	73,	34,	14

Lesson 5 Number 74

	a	b	c	d	e
# 1	74,	64,	74,	44,	24
# 2	74,	63,	74,	43,	23
# 3	74,	62,	74,	42,	22
# 4	74,	61,	74,	41,	21
# 5	74,	60,	74,	40,	20
# 6	74,	59,	74,	39,	19
# 7	74,	58,	74,	38,	18
# 8	74,	57,	74,	37,	17
# 9	74,	56,	74,	36,	16
# 10	74,	55,	74,	35,	15

Lesson 6 Number 75

	a	b	c	d	e
# 1	75,	65,	75,	45,	25
# 2	75,	64,	75,	44,	24
# 3	75,	63,	75,	43,	23
# 4	75,	62,	75,	42,	22
# 5	75,	61,	75,	41,	21
# 6	75,	60,	75,	40,	20
# 7	75,	59,	75,	39,	19
# 8	75,	58,	75,	38,	18
# 9	75,	57,	75,	37,	17
# 10	75,	56,	75,	36,	16

Summary 3 Answers Chapter 3

Lesson 7 Number 76

	a, b, c, d, e
# 1	76, 66, 76, 46, 26
# 2	76, 65, 76, 45, 25
# 3	76, 64, 76, 44, 24
# 4	76, 63, 76, 43, 23
# 5	76, 62, 76, 42, 22
# 6	76, 61, 76, 41, 21
# 7	76, 60, 76, 40, 20
# 8	76, 59, 76, 39, 19
# 9	76, 58, 76, 38, 18
# 10	76, 57, 76, 37, 17

Lesson 8 Number 77

	a, b, c, d, e
# 1	77, 67, 77, 47, 27
# 2	77, 66, 77, 46, 26
# 3	77, 65, 77, 45, 25
# 4	77, 64, 77, 44, 24
# 5	77, 63, 77, 43, 23
# 6	77, 62, 77, 42, 22
# 7	77, 61, 77, 41, 21
# 8	77, 60, 77, 40, 20
# 9	77, 59, 77, 39, 19
# 10	77, 58, 77, 38, 18

Lesson 9 Number 78

	a, b, c, d, e
# 1	78, 68, 78, 48, 28
# 2	78, 67, 78, 47, 27
# 3	78, 66, 78, 46, 26
# 4	78, 65, 78, 45, 25
# 5	78, 64, 78, 44, 24
# 6	78, 63, 78, 43, 23
# 7	78, 62, 78, 42, 22
# 8	78, 61, 78, 41, 21
# 9	78, 60, 78, 40, 20
# 10	78, 59, 78, 39, 19

Lesson 10 Number 79

	a, b, c, d, e
# 1	79, 69, 79, 49, 29
# 2	79, 68, 79, 48, 28
# 3	79, 67, 79, 47, 27
# 4	79, 66, 79, 46, 26
# 5	79, 65, 79, 45, 25
# 6	79, 64, 79, 44, 24
# 7	79, 63, 79, 43, 23
# 8	79, 62, 79, 42, 22
# 9	79, 61, 79, 41, 21
# 10	79, 60, 79, 40, 20

Lesson 11 Test 3 A

1)	a)	70	b)	35
2)	a)	70	b)	36
3)	a)	71	b)	37
4)	a)	71	b)	38
5)	a)	72	b)	39
6)	a)	72	b)	25
7)	a)	73	b)	26
8)	a)	73	b)	27
9)	a)	74	b)	28
10)	a)	74	b)	29

Lesson 11 Test 3 B

1)	a)	75	b)	25
2)	a)	75	b)	26
3)	a)	76	b)	27
4)	a)	76	b)	28
5)	a)	77	b)	29
6)	a)	77	b)	35
7)	a)	78	b)	36
8)	a)	78	b)	37
9)	a)	79	b)	38
10)	a)	79	b)	39

Summary 3 Answers Chapter 3

Lesson 12

Review 3 A

# 1	90	# 6	35	# 11	91	# 16	19
# 2	94	# 7	58	# 12	93	# 17	28
# 3	98	# 8	24	# 13	95	# 18	45
# 4	92	# 9	48	# 14	97	# 19	39
# 5	96	#10	40	# 15	90	# 20	56

Review 3 B

# 1	91	# 6	34	# 11	90	# 16	62
# 2	95	# 7	57	# 12	92	# 17	59
# 3	99	# 8	25	# 13	94	# 18	66
# 4	93	# 9	37	# 14	96	# 19	49
# 5	97	# 10	69	# 15	95	# 20	56

Lesson 13

Final Exam 3 A

# 1	90	# 11	95
# 2	91	# 12	96
# 3	92	# 13	97
# 4	93	# 14	98
# 5	94	# 15	99
# 6	59	# 16	46
# 7	22	# 17	39
# 8	79	# 18	68
# 9	24	# 19	19
# 10	59	# 20	80

Final Exam 3 B

# 1	90	# 11	95
# 2	91	# 12	96
# 3	92	# 13	97
# 4	93	# 14	98
# 5	94	# 15	99
# 6	49	# 16	36
# 7	32	# 17	29
# 8	69	# 18	58
# 9	34	# 19	29
# 10	49	# 20	70

Math Help

for College

Summary 4

Summary 4 Pretest 4
Evaluate

 # 1 $8 \times 5 \times 6$

 # 2 $9 \times 24 \div 8$

 # 3 $56 \div 8 \times 4$

 # 4 $640 \div 8 \div 4$

 # 5 $109 + 9 \times 9$

 # 6 $60 \times 8 - 80$

 # 7 $240 - 8 \times 5$

 # 8 $720 \div 9 - 20$

 # 9 $333 + 444 \div 4$

 # 10 $360 - 360 \div 6$

Summary 4

Chapter 1 Three-Digit Numbers

Summary 4 Chapter 1 Contents
Three-digit numbers
Lesson 1 – 10
Basic Problems

Work with lesson 1, read rules and examples, 2) Solve the problems from contents, 3) Check answers in the end of chapter 1, 4) Work with the next lessons 2, 3, 4, 5, ..., 5) Solve Test 1 A, 6) Grading Scale: A =10 points, B=9points, C=8 points, D=7 points, F=0-6 points, 7) If you solve 10 problems correctly, you get an A, 8) If have an A, you can move on to next chapter, 9) If your grade is lower than an A, work with the mistakes, then solve Test 1B, 10) When you finish all of the chapters, solve Final Exam in last chapter. Lesson = L

L1 One-digit and two-digit numbers. Solve # 11–20. Pretest #1-# 10
L2 Addition, Subtraction Numbers. Solve Review 1 A, Review 1 B # 5, 10

L3 Two operations with numbers. Solve Review 1 A, Review 1 B # 5, 10
L4 Addition, Subtraction three-digit numbers. Solve # 4, # 8

L5 Numbers from 100 to 199. Solve # 4, # 8 (a, b, c, d, e)
L6 Numbers from 200 to 299. Solve # 4, # 8 (a, b, c, d, e)

L7 Numbers from 300 to 399. Solve # 4, # 8 (a, b, c, d, e)
L8 Addition and Subtraction numbers.
Solve the problems # 5, # 10, # 15, # 20 (a, b, c)

L9 Subtraction and Addition numbers.
Solve the problems # 5, # 10, # 15, # 20 (a, b, c)
L10 Test 1 A # 1, # 2, # 3, # 4, # 5, # 6, # 7, # 8, # 9, # 10 (a)
Test 1 B # 1, # 2, # 3, # 4, # 5, # 6, # 7, # 8, # 9, # 10 (a)

Answers Chapter 1

Summary 4 Chapter 1 Lesson 1
One-digit and two-digit numbers

One-digit numbers
Count forward from 1 to 9 1, 2, 3, 4, 5, 6, 7, 8, 9
Count backwards from 9 to 1 9, 8, 7, 6, 5, 4, 3, 2, 1
Equal = Less than < Greater than >
Compare one-digit numbers (put <, or >, or =)
5 = 5 0 < 1 5 < 8 6 > 4 9 > 8
Two-digit numbers
Digits from 0 to 9 0, 1, 2, 3, 4, 5, 6, 7, 8, 9
One-digit numbers have one digit. This is numbers from 0 to 9
Two-digit numbers have two digits. This is numbers from 10 to 99
Digit from the left called tens. Digit from the right called ones
Comparison one-digit and two-digit numbers
One-digit numbers always less than two-digit numbers 9 < 11
Two-digit numbers always greater than one-digit numbers 19 > 8
To compare two-digit numbers:
Compare the tens, then compare the ones 12 < 18 18 > 12
Two operations with numbers. Numbers 80 – 99
Three numbers can add in any order: 11+14+19=11+19+14=30+14=44
To do two operations with numbers, work from left to right
To subtract three numbers, work from left to right:
18 – 2 – 8 = 16 – 8 = 8
To add and subtract numbers, work from left to right:
18 + 8 - 2 = 26 - 2 = 24
To subtract and add, work from left to right: 18 - 9 + 8 = 9 + 8 = 17
Add and Subtract Numbers: 68 – 29 = 68 – 20 – 9 = 48 – 9 = 39
39 + 29 =30 + 9 + 20 + 9 = 30 + 20 + 9 + 9 = 50 + 18 = 68
Two numbers can add in any order. The sum will be the same.
2 + 3 = 5 3 + 2 = 5
The sum of a number and zero is the same number:
5 + 0 = 5 0 + 5 = 5
Difference any number and zero is the same number:
5 – 0 = 5
Difference two the same number is zero: 5 – 5 = 0
To check addition use operation subtraction
2 + 3 = 5 Check 5 – 2 = 3 or 5 – 3 = 2
To check subtraction use operation addition and subtraction
5 – 2 = 3 Check 2 + 3 = 5 or 5 – 3 = 2

Summary 4 Chapter 1 Lesson 1
Addition and Subtraction Numbers
Addition

First number in addition called the addend	2
Second number in addition called the addend	3
Result of addition called the sum	5

Subtraction

First number in subtraction called the minuend	5
Second number in subtraction called the subtrahend	2
Result of subtraction called the difference	3

Add or Subtract Numbers in row

# 1	18+12=10+8+10+2=20+10=30	# 6	30–22=30-20-2=10-2=8
# 2	22 + 9	# 7	31 – 12
# 3	28 + 6	# 8	34 – 27
# 4	18 + 17	# 9	35 – 19
# 5	29 + 7	# 10	36 – 9

Add or Subtract numbers in column

#11 12	# 12 26	# 13 26	# 14 28	# 15 35
+19	+ 17	+ 19	+19	+14

# 16 41	# 17 43	# 18 45	# 19 47	# 20 49
-28	-29	-16	- 19	- 15

Summary 4 Chapter 1 Lesson 2

Put plus or minus with numbers

Review 1 A

Put + or – between numbers that each problem was correct

20 ? 60 = 80 20 + 60 = 80 60 ? 20 = 40 60 - 20 = 40

	a	b	c	d
# 1	20 ? 40=60	30 ? 30=60	40 ? 20=60	50 ? 10=60
# 2	60 ? 10=50	70 ? 10=60	80 ? 10=70	90 ? 10=80
# 3	21 ? 42=63	31 ? 32=63	41 ? 22=63	51 ? 12=63
# 4	61 ? 12=49	71 ? 12=59	81 ? 12=69	91 ? 12=79
# 5	22 ? 44=66	32 ? 34=66	42 ? 24=66	52 ? 14=66

Put + or – between numbers that each problem was correct

20 ? 70 = 90 20 + 70 = 90 90 ? 20 = 70 90 - 20 = 70

	a	b	c	d
# 6	25 ? 41=66	35 ? 31=66	45 ? 21=66	55 ? 11=66
# 7	65 ? 11=54	75 ? 11=64	85 ? 11=74	95 ? 11=84
# 8	26 ? 43=69	36 ? 33=69	46 ? 23=69	56 ? 13=69
# 9	66 ? 13=53	76 ? 13=63	86 ? 13=73	96 ? 13=83
# 10	27 ? 45=72	37 ? 35=72	47 ? 25=72	57 ? 15=72

Rewrite, Read, Memorize Numbers

Numbers from 1 to 10 by 1	1, 2, 3, 4, 5, 6, 7, 8, 9, 10
Numbers from 2 to 20 by 2	2, 4, 6, 8, 10, 12, 14, 16, 18, 20
Numbers from 3 to 30 by 3	3, 6, 9, 12, 15, 18, 21, 24, 27, 30
Numbers from 4 to 40 by 4	4, 8, 12, 16, 20, 24, 28, 32, 36, 40
Numbers from 5 to 50 by 5	5, 10, 15, 20, 25, 30, 35, 40, 45, 50

Review 1 B

Addition the same numbers

48 + 48 = 40 + 8 +40 + 8 = 40 + 40 + 8 + 8 = 80 +16 = 96

	a	b	c	d
# 1	11+11	21+21	31+31	41+41
# 2	12+12	22+22	32+32	42+42
# 3	13+13	23+23	33+33	43+43
# 4	14+14	24+24	34+34	44+44
# 5	15+15	25+25	35+35	45+45
# 6	21+21	31+31	41+41	51+51
# 7	22+22	32+32	42+42	52+52
# 8	23+23	33+33	43+43	53+53
# 9	24+24	34+34	44+44	54+54
# 10	25+25	35+35	45+45	55+55

Summary 4 Chapter 1 Lesson 3

Add three numbers

Three numbers add in any order: 11+14+19=11+19+14=30+14=44

11 + 14 + 19 = 25 + 19 = 44   11 + 14 + 19 = 11 + 33 = 44

Review 1 A Add numbers

	a	b	c	d
# 1	20 + 0 + 9	30 + 0 + 9	40 + 0 + 9	50 + 0 + 9
# 2	21 + 1 + 8	31 + 1 + 8	41 + 1 + 8	51 + 1 + 8
# 3	22 + 2 + 7	32 + 2 + 7	42 + 2 + 7	52 + 2 + 7
# 4	23 + 3 + 6	33 + 3 + 6	43 + 3 + 6	53 + 3 + 6
# 5	24 + 4 + 5	34 + 4 + 5	44 + 4 + 5	54 + 4 + 5

Subtract three numbers

To subtract three numbers, work from left to right

19 – 1 – 9 = 18 – 9 = 9   18 – 2 – 8 = 16 – 8 = 8   18 – 8 – 2 = 10 – 2 = 8

Subtract numbers

	a	b	c	d
# 6	60 – 9 – 0	70 – 9 – 0	80 – 9 – 0	90 – 9 – 0
# 7	61 – 8 – 1	71 – 8 – 1	81 – 8 – 1	91 – 8 – 1
# 8	62 – 7 – 2	72 – 7 – 2	82 - 7 – 2	92 – 7 – 2
# 9	63 – 6 – 3	73 – 6 – 3	83 – 6 – 3	93 – 6 – 3
# 10	64 – 5 – 4	74 – 5 – 4	84 – 5 – 4	94 – 5 – 4

Two operations with numbers

To do two operations with numbers work from left to right

19 + 7 - 1 = 26 - 1 = 25   18 + 8 - 2 = 26 - 2 = 24

Add and Subtract numbers. Review 1 B

	a	b	c	d
# 1	20 + 9 – 0	30 + 9 – 0	40 + 9 – 0	50 + 9 - 0
# 2	21 + 8 – 1	31 + 8 – 1	41 + 8 – 1	51 + 8 - 1
# 3	22 + 7 – 2	32 + 7 – 2	42 + 7 – 2	52 + 7 - 2
# 4	23 + 6 – 3	33 + 6 – 3	43 + 6 – 3	53 + 6 - 3
# 5	24 + 5 – 4	34 + 5 – 4	44 + 5 – 4	54 + 5 - 4

To do two operations with numbers work from left to right

19 - 8 + 9 = 11 + 9 = 20   18 - 9 + 8 = 9 + 8 = 17

Subtract and Add Numbers

	a	b	c	d
# 6	20 - 9 + 0	30 - 9 + 0	40 - 9 + 0	50 - 9 + 0
# 7	21 - 8 + 1	31 - 8 + 1	41 - 8 + 1	51 - 8 + 1
# 8	22 - 7 + 2	32 - 7 + 2	42 - 7 + 2	52 - 7 + 2
# 9	23 - 6 + 3	33 - 6 + 3	43 - 6 + 3	53 - 6 + 3
# 10	24 - 5 + 4	34 - 5 + 4	44 - 5 + 4	54 - 5 + 4

Summary 4 Chapter 1 Lesson 4
Addition and Subtraction
Three-digit numbers

Three-digit numbers content three digits
This is numbers 100, 101, 102, ... 999
Least three-digit number: 100
Greatest three-digit number: 999
Digits for three-digit numbers from right to the left:
Ones, tens, and hundreds 238 = 200 + 30 + 8
One-digit numbers less than two or three-digit numbers
Three-digit numbers greater than one or two-digit numbers

To compare numbers, compare first hundred, after that tens, then ones
$423 < 456$ $789 > 786$
Two numbers add in any order. Sum will be the same.
200 + 300 = 500 300 + 200 = 500
The sum of a number and zero is the same number
500 + 0 = 500 0 + 500 = 500
Difference any number and zero is the same number 500 – 0 = 500
Difference two the same number is zero 500 – 500 = 0

Adding three-digit numbers: from right to left
Add the ones. Add the tens. Add the hundreds.
Subtracting three-digit numbers: from right to left
Subtract the ones. Subtract the tens. Subtract the hundreds.

		1 1	7 14 14
246	678	154	8 5 4
+ 351	- 432	+789	- 2 8 6
597	246	943	5 6 8

Summary 4 Chapter 1 Lesson 4

Add three-digit numbers

	a	b	c	d	e
# 1	121 +594	132 +616	143 +715	154 +495	465 +484
# 2	252 +732	276 +660	312 +629	408 +588	561 +348
# 3	273 +598	286 +589	299 +592	377 +559	429 +546
# 4	294 +532	378 +476	392 +469	322 + 448	336 +439

Subtract three-digit numbers

	a	b	c	d	e
# 5	966 -357	945 -378	924 -399	903 -504	882 -525
# 6	989 -598	943 -575	827 -499	862 -483	815 -437
# 7	999 -108	955 -359	891 -297	837 -248	784 -189
# 8	957 -609	899 -493	841 -495	763 -377	718 -319

Summary 4 Chapter 1 Lesson 5
Sum from 100 to 199
Add numbers in column

	a	b	c	d	e
# 1	95 +22	96 +24	97 +26	98 +28	99 +29
# 2	38 +97	37 +96	36 +95	35 +99	34 +98
# 3	98 +43	97 +45	96 +47	95 + 49	97 +48
# 4	58 +97	57 +95	56 +98	55 +96	54 +99

Subtract numbers in column

	a	b	c	d	e
# 5	121 - 97	120 - 94	123 - 96	126 - 98	128 - 99
# 6	135 - 97	133 - 96	131 - 95	134 - 99	132 - 98
# 7	141 - 98	142 - 97	143 - 96	144 - 95	145 - 97
# 8	155 - 97	152 - 95	154 - 98	151 - 96	153 - 99

Summary 4 Chapter 1 Lesson 6
Sum from 200 to 299
Add numbers in column

	a	b	c	d	e
# 1	195 +22	196 +24	197 +26	198 +28	199 +29
# 2	138 +97	137 +96	136 +95	135 +99	134 +98
# 3	198 +43	197 +45	196 +47	195 +49	197 +48
# 4	158 +97	157 +95	156 +98	155 +96	154 +99

Subtract numbers in column

	a	b	c	d	e
# 5	221 - 97	220 - 94	223 - 96	226 - 98	228 - 99
# 6	235 - 97	233 - 96	231 - 95	234 - 99	232 - 98
# 7	241 - 98	242 - 97	243 - 96	244 - 95	245 - 97
# 8	255 - 97	252 - 95	254 - 98	251 - 96	253 - 99

Summary 4 Chapter 1 Lesson 7
Sum from 300 to 399
Add numbers in column

	a	b	c	d	e
# 1	195 +122	196 +124	197 +126	198 +128	199 +129
# 2	138 +197	137 +196	136 +195	135 +199	134 +198
# 3	198 +143	197 +145	196 +147	195 +149	197 +148
# 4	158 +197	157 +195	156 +198	155 +196	154 +199

Subtract numbers in column

	a	b	c	d	e
# 5	321 - 97	320 - 94	323 - 96	326 - 98	328 - 99
# 6	335 - 97	333 - 96	331 - 95	334 - 99	332 - 98
# 7	341 - 98	342 - 97	343 - 96	344 - 95	345 - 97
# 8	355 - 97	352 - 95	354 - 98	351 - 96	353 - 99

Summary 4 Chapter 1 Lesson 8

Addition numbers

To add numbers, work in any order: 410 + 140 + 290 = 550 + 290 = 840

410 + 140 + 290 = 410 + 430 = 840, 410 + 140 + 290=410 + 290 + 140=840

Review 1 A Evaluate

	a	b	c
# 1	10 + 90 + 10	20 + 80 + 20	30 + 70 + 30
# 2	40 + 60 + 40	50 + 50 + 50	60 + 40 + 60
# 3	70 + 30 + 70	80 + 20 + 80	90 + 10 + 90
# 4	100 + 0 +100	110 + 90 + 10	120 + 80 + 20
# 5	130 + 70 + 30	140 + 60 + 40	150 + 50 + 50

	a	b	c
# 6	310 + 90 + 290	320 + 80 + 280	330 + 70 + 270
# 7	340 + 60 + 260	350 + 50 + 250	360 + 40 + 240
# 8	370 + 30 + 230	380 + 20 + 220	390 + 10 + 210
# 9	400 + 20 + 300	410 + 30 + 290	420 + 40 + 280
# 10	430 + 50 + 270	440 + 60 + 260	450 + 70 + 250

Addition and Subtraction numbers

To add and subtract numbers work from left to right

410 + 140 - 110 = 550 - 110 = 440

Review 1 B

	a	b	c
# 11	10 + 90 - 0	20 + 80 - 10	30 + 70 - 20
# 12	40 + 60 - 30	50 + 50 - 40	60 + 40 - 50
# 13	70 + 30 - 60	80 + 20 - 70	90 + 10 - 80
# 14	100 + 100 - 0	110 + 90 - 10	120 + 80 - 20
# 15	130 + 70 - 30	140 + 60 - 40	150 + 50 - 50

	a	b	c
# 16	310 + 90 - 10	320 + 80 - 20	330 + 70 - 30
# 17	340 + 60 - 40	350 + 50 - 50	360 + 40 - 60
# 18	370 + 30 - 70	380 + 20 - 80	390 + 10 - 90
# 19	400 + 300 - 0	410 + 290 - 10	420 + 280 - 20
# 20	430 + 270 - 30	440 + 260 - 40	450 + 250 - 50

Add numbers in column.

$$\begin{array}{r} 220 \\ +\,150 \\ \underline{380} \\ 750 \end{array}$$

Add and subtract numbers in column

410 + 140 - 110 = 440

1) $$\begin{array}{r} 410 \\ \underline{+\,140} \\ 550 \end{array}$$

2) $$\begin{array}{r} 550 \\ \underline{-\,110} \\ 440 \end{array}$$

Summary 4 Chapter 1 Lesson 9

Subtraction and Addition numbers

To subtract and add numbers work from left to right

410 - 110 + 140 = 300 + 140 = 440

Review 1 A Evaluate

	a	b	c
# 1	10 – 0 + 90	20 – 10 + 80	30 – 20 + 70
# 2	40 – 30 + 60	50 – 40 + 50	60 – 50 + 40
# 3	70 – 60 + 30	80 – 70 + 20	90 – 80 + 10
# 4	100 – 0 + 100	110 – 10 + 90	120 – 20 + 80
# 5	130 – 30 + 70	140 – 40 + 60	150 - 50 + 50

	a	b	c
# 6	310 – 10 + 90	320 – 20 + 80	330 – 30 + 70
# 7	340 – 40 + 60	350 - 50 + 50	360 – 60 + 40
# 8	370 – 70 + 30	380 – 80 + 20	390 – 90 + 10
# 9	400 – 0 + 300	410 – 10 + 290	420 – 20 + 280
# 10	430 – 30 + 270	440 – 40 + 260	450 – 50 + 250

Subtraction numbers

To subtract numbers, work from left to right

900 – 200 – 100 = 700 – 100 = 600

Review 1 B

	a	b	c
# 11	400 – 10 - 100	410 – 20 - 110	420 – 30 - 120
# 12	430 – 40 - 130	440 – 50 – 140	450 – 60 - 150
# 13	460 – 70 – 160	470 - 80 – 170	480 – 90 - 180
# 14	490 – 100 – 190	500 – 110 – 200	510 – 120 - 210
# 15	520 – 130 – 220	530 – 140 – 230	540 – 150 - 240

	a	b	c
# 16	700 – 10 – 400	710 – 20 – 410	720 – 30 - 420
# 17	730 – 40 – 430	740 – 50 – 440	750 – 60 - 450
# 18	760 – 70 – 460	770 – 80 – 470	780 – 90 - 480
# 19	790 – 100 – 490	800 – 110 – 500	810 – 120 - 510
# 20	820 – 130 – 520	830 – 140 – 530	840 – 150 - 540

Subtract and add numbers in column. Subtract numbers in column

410 - 110 + 140 = 440

$$\begin{array}{r} 410 \\ -\ 110 \\ \hline 300 \end{array} \qquad \begin{array}{r} 300 \\ +\ 140 \\ \hline 440 \end{array}$$

1) and 2) above.

900 – 200 – 100 = 600

$$\begin{array}{r} 900 \\ -\ 200 \\ \hline 700 \end{array} \qquad \begin{array}{r} 700 \\ -\ 100 \\ \hline 600 \end{array}$$

Summary 4 Chapter 1 Lesson 10
Test 1 A
Add numbers in column

	a	b
# 1	196 +189	194 +186

Subtract numbers in column

	a	b
# 2	385 - 96	380 - 94

Evaluate in row or in column

	a	b
# 3	250 + 50 + 50	260 + 40 + 60
# 4	550 + 170 + 150	560 + 180 + 140
# 5	250 + 50 - 50	260 + 40 - 60
# 6	550 + 150 - 150	560 + 140 - 160
# 7	250 - 50 + 50	260 – 60 + 40
# 8	550 - 150 + 150	560 – 160 + 140
# 9	640 – 250 – 340	650 – 260 – 350
# 10	940 – 250 – 640	950 – 260 – 650

Summary 4 Chapter 1 Lesson 10
Test 1 B

Add numbers in column

	a	b
# 1	198 +197	197 +195

Subtract numbers in column

	a	b
# 2	395 - 97	392 - 95

Evaluate in row or in column

	a	b
# 3	280 + 20 + 80	290 + 10 + 90
# 4	580 + 200 + 120	590 + 210 + 110
# 5	280 + 20 - 80	290 + 10 - 90
# 6	580 + 120 - 180	590 + 110 - 190
# 7	280 – 80 + 20	290 – 90 + 10
# 8	580 – 180 + 120	590 – 190 + 110
# 9	670 – 280 – 370	680 – 290 – 380
# 10	970 – 280 – 670	980 – 290 – 680

Summary 4 Answers Chapter 1

Pretest 4 1) 240, 2) 27, 3) 28, 4) 20, 5) 190, 6) 400, 7) 200, 8) 60, 9) 444, 10) 300

Lesson 1

# 1	30	# 6	8	# 11	31	# 16	13
# 2	31	# 7	19	# 12	43	# 17	14
# 3	34	# 8	7	# 13	45	# 18	29
# 4	35	# 9	16	# 14	47	# 19	28
# 5	36	# 10	27	# 15	49	# 20	34

Lesson 2 Review 1 A

	a	b	c	d		a	b	c	d
# 1	+	+	+	+	# 6	+	+	+	+
# 2	-	-	-	-	# 7	-	-	-	-
# 3	+	+	+	+	# 8	+	+	+	+
# 4	-	-	-	-	# 9	-	-	-	-
# 5	+	+	+	+	# 10	+	+	+	+

Lesson 2 Review 1 B

	a b c d		a b c d
# 1	22, 42, 62, 82	# 6	42, 62, 82, 102
# 2	24, 44, 64, 84	# 7	44, 64, 84, 104
# 3	26, 46, 66, 86	# 8	46, 66, 86, 106
# 4	28, 48, 68, 88	# 9	48, 68, 88, 108
# 5	30, 50, 70, 90	# 10	50, 70, 90, 110

Lesson 3 Review 1 A

	a b c d		a b c d
# 1	29, 39, 49, 59	# 6	51, 61, 71, 81
# 2	30, 40, 50, 60	# 7	52, 62, 72, 82
# 3	31, 41, 51, 61	# 8	53, 63, 73, 83
# 4	32, 42, 52, 62	# 9	54, 64, 74, 84
# 5	33, 43, 53, 63	# 10	55, 65, 75, 85

Lesson 3 Review 1 B

	a b c d		a b c d
# 1	29, 39, 49, 59	# 6	11, 21, 31, 41
# 2	28, 38, 48, 58	# 7	14, 24, 34, 44
# 3	27, 37, 47, 57	# 8	17, 27, 37, 47
# 4	26, 36, 46, 56	# 9	20, 30, 40, 50
# 5	25, 35, 45, 55	# 10	23, 33, 43, 53

Summary 4 Answers Chapter 1

Lesson 4

	a	b	c	d	e
# 1	715,	748,	858,	649,	949
# 2	984,	936,	941,	996,	909
# 3	871,	875,	891,	936,	975
# 4	826,	854,	861,	770,	775
# 5	609,	567,	525,	399,	357
# 6	391,	368,	328,	379,	378
# 7	891,	596,	594,	589,	595
# 8	348,	406,	346,	386,	399

Lesson 5

	a	b	c	d	e
# 1	117,	120,	123,	126,	128
# 2	135,	133,	131,	134,	132
# 3	141,	142,	143,	144,	145
# 4	155,	152,	154,	151,	153
# 5	24,	26,	27,	28,	29
# 6	38,	37,	36,	35,	34
# 7	43,	45,	47,	49,	48
# 8	58,	57,	56,	55,	54

Lesson 6

	a	b	c	d	e
# 1	217,	220,	223,	226,	228
# 2	235,	233,	231,	234,	232
# 3	241,	242,	243,	244,	245
# 4	255,	252,	254,	251,	253
# 5	124,	126,	127,	128,	129
# 6	138,	137,	136,	135,	134
# 7	143,	145,	147,	149,	148
# 8	158,	157,	156,	155,	154

Lesson 7

	a	b	c	d	e
# 1	317,	320,	323,	326,	328
# 2	335,	333,	331,	334,	332
# 3	341,	342,	343,	344,	345
# 4	355,	352,	354,	351,	353
# 5	224,	226,	227,	228,	229
# 6	238,	237,	236,	235,	234
# 7	243,	245,	247,	249,	248
# 8	258,	257,	256,	255,	254

Lesson 8 Review 1 A

	a	b	c
# 1	110	120	130
# 2	140	150	160
# 3	170	180	190
# 4	200	210	220
# 5	230	240	250
# 6	690	680	670
# 7	660	650	640
# 8	630	620	610
# 9	720	730	740
# 10	750	760	770

Review 1 B

	a	b	c
# 11	100	90	80
# 12	70	60	50
# 13	40	30	20
# 14	200	190	180
# 15	170	160	150
# 16	390	380	370
# 17	360	350	340
# 18	330	320	310
# 19	700	690	680
# 20	670	660	650

Summary 4 Answers Chapter 1

Lesson 9 Review 1 A

	a	b	c
# 1	100	90	80
# 2	70	60	50
# 3	40	30	20
# 4	200	190	180
# 5	170	160	150
# 6	390	380	370
# 7	360	350	340
# 8	330	320	310
# 9	700	690	680
# 10	670	660	650

Review 1 B

	a	b	c
# 11	290	280	270
# 12	260	250	240
# 13	230	220	210
# 14	200	190	180
# 15	170	160	150
# 16	290	280	270
# 17	260	250	240
# 18	230	220	210
# 19	200	190	180
# 20	170	160	150

Lesson 10

Test 1 A

	a	b
# 1	385	380
# 2	289	286
# 3	350	360
# 4	870	880
# 5	250	240
# 6	550	540
# 7	250	240
# 8	550	540
# 9	50	40
# 10	50	40

Test 1 B

	a	b
# 1	395	392
# 2	298	297
# 3	380	390
# 4	900	910
# 5	220	210
# 6	520	510
# 7	220	210
# 8	520	510
# 9	20	10
# 10	20	10

Summary 4

Chapter 2

Multiplication

Numbers 1

Summary 4 Chapter 2 Contents
Multiplication Numbers
Lesson 1 – 14
Basic Problems

Work with lesson 1, read rules and examples, 2) Solve the problems from contents, 3) Check answers in the end of chapter 1, 4) Work with the next lessons 2, 3, 4, 5, …, 5) Solve Test 1 A, 6) Grading Scale: A =10 points, B=9points, C=8 points, D=7 points, F=0-6 points, 7) If you solve 10 problems correctly, you get an A, 8) If have an A, you can move on to next chapter, 9) If your grade is lower than an A, work with the mistakes, then solve Test 1B, 10) When you finish all of the chapters, solve Final Exam in last chapter. Lesson = L

L1Multiplication Numbers. Solve the problems # 10 (a, b), # 16 (a, b)
L2Complete and Memorize Table of Multiplication # 10, 20(a, b, c, d. e)
L3Multiplication Numbers from 10 to 19. Solve the problems#4,# 8(a-e)
L4Multiplication Numbers from 20 to 29. Solve the problems#4,#8 (a-e)
L5 Multiplication Numbers from 30 to 39. Solve the problems#4,#8(a-e)
L6 Multiplication Numbers from 40 to 49. Solve the problems#4,#8(a-e)
L7 Multiplication Numbers from 50 to 59. Solve the problems#4,#8 a-e)
L8 Multiplication Numbers from 60 to 69. Solve the problems#4,#8 (a-e)
L9 Multiplication Numbers from 70 to 79. Solve the problems#4,#8(a-e)
L10 Multiplication Numbers from 80 to 89.Solve the problems#4,#8(a-e)
L11Multiplication Numbers from 90 to 99. Solve the problems#4,#8(a-e)
L12 Multiplication and Addition or Subtraction numbers.
Solve the problems Review 1 A # 5, # 10 (a-d), Review 1 B # 5, 10 (a-d)
L13 Read summary 1 – 2. Multiply numbers.
Review 2 A #1, 3, 5, 7, 9, 11, 13, 15, 17, 19
Review 2 B #1, 3, 5, 7, 9, 11, 13, 15, 17, 19
L14 Test 2 A, Test 2 B for 50 minutes
Solve Test 2 A #1, #3, #5, #7, #9, #11, #13, #15, #17, #19
Solve Test 2 B #1, 3, 5, 7, 9. 11, 13, 15, 17, 19
Answers Chapter 2

Summary 4 Chapter 2 Lesson 1
Multiplication one-digit numbers

Third operation with numbers is multiplication × or ·

Multiplication is the operation of repeated addition

$2\times 3 = 2+2+2 = 6$ or $2\cdot 3 = 2+2+2 = 6$

First number in multiplication called the factor	**2**
Second number in multiplication called the factor	**3**
Result of multiplication called the product	**6**

Properties of multiplication

Two numbers can multiply in any order. The product will be the same.

$2\times 3 = 2+2+2 = 6$ $\quad$ $3\times 2 = 3+3 = 6$

The product of a number and zero is zero: $5\times 0 = 0$ $\quad$ $0\times 5 = 0$

The product of a number and one is the same number:

$5\times 1 = 5$ $\quad$ $1\times 5 = 5$

Find the product: <u>$4\times 8 = 8\times 4 = 8+8+8+8 = 16+16 = 32$</u>

<u>This exercises help you to memorize table of multiplication</u>

Check please:	a	b
# 1	$0\times 2 = 0+0 = 0$	$0\times 3 = 0+0+0 = 0$
# 2	$1\times 2 = 1+1 = 2$	$1\times 3 = 1+1+1 = 3$
# 3	$2\times 2 = 2+2 = 4$	$2\times 3 = 2+2+2 = 6$
# 4	$3\times 2 = 3+3 = 6$	$3\times 3 = 3+3+3 = 9$
# 5	$4\times 2 = 4+4 = 8$	$4\times 3 = 4+4+4 = 12$
# 6	$5\times 2 = 5+5 = 10$	$5\times 3 = 5+5+5 = 15$
# 7	$6\times 2 = 6+6 = 12$	$6\times 3 = 6+6+6 = 18$
# 8	$7\times 2 = 7+7 = 14$	$7\times 3 = 7+7+7 = 21$
# 9	$8\times 2 = 8+8 = 16$	$8\times 3 = 8+8+8 = 24$
# 10	$9\times 2 = 9+9 = 18$	$9\times 3 = 9+9+9 = 27$

	a	b		a	b
# 11	$0\times 4 = 0$	$0\times 5 = 0$	# 21	$6\times 0 = 0$	$7\times 0 = 0$
# 12	$1\times 4 = 4$	$1\times 5 = 5$	# 22	$6\times 1 = 6$	$7\times 1 = 7$
# 13	$2\times 4 = 8$	$2\times 5 = 10$	# 23	$6\times 2 = 12$	$7\times 2 = 14$
# 14	$3\times 4 = 12$	$3\times 5 = 15$	# 24	$6\times 3 = 18$	$7\times 3 = 21$
# 15	$4\times 4 = 16$	$4\times 5 = 20$	# 25	$6\times 4 = 24$	$7\times 4 = 28$
# 16	$5\times 4 = 20$	$5\times 5 = 25$	# 26	$6\times 5 = 30$	$7\times 5 = 35$
# 17	$6\times 4 = 24$	$6\times 5 = 30$	# 27	$6\times 6 = 36$	$7\times 6 = 42$
# 18	$7\times 4 = 28$	$7\times 5 = 35$	# 28	$6\times 7 = 42$	$7\times 7 = 49$
# 19	$8\times 4 = 32$	$8\times 5 = 40$	# 29	$6\times 8 = 48$	$7\times 8 = 56$
# 20	$9\times 4 = 36$	$9\times 5 = 45$	# 30	$6\times 9 = 54$	$7\times 9 = 63$

Summary 4 Chapter 2 Lesson 2

Multiply the numbers and memorize the result:

$2\times3=2+2+2=6$ $3\times2=3+3=6$ $3\times9=9\times3=9+9+9=18+9=27$

	a	b	c	d
# 1	2×1	3×1	4×1	5×1
# 2	2×2	3×2	4×2	5×2
# 3	2×3	3×3	4×3	5×3
# 4	2×4	3×4	4×4	5×4
# 5	2×5	3×5	4×5	5×5
# 6	2×6	3×6	4×6	5×6
# 7	2×7	3×7	4×7	5×7
# 8	2×8	3×8	4×8	5×8
# 9	2×9	3×9	4×9	5×9
# 10	2×10	3×10	4×10	5×10
#11	6×1	7×1	8×1	9×1
# 12	6×2	7×2	8×2	9×2
# 13	6×3	7×3	8×3	9×3
# 14	6×4	7×4	8×4	9×4
# 15	6×5	7×5	8×5	9×5
# 16	6×6	7×6	8×6	9×6
# 17	6×7	7×7	8×7	9×7
# 18	6×8	7×8	8×8	9× 8
# 19	6×9	7×9	8×9	9×9
# 20	6×10	7×10	8×10	9×10

Table of multiplication

Memorize the Table of Multiplication. Number 5 from the left column multiply by number 4 from first row to get, answer 20

	×	2	3	4	5	6	7	8	9
# 1	1	2	3	4	5	6	7	8	9
# 2	2	4	6	8	10	12	14	16	18
# 3	3	6	9	12	15	18	21	24	27
# 4	4	8	12	16	20	24	28	32	36
# 5	5	10	15	20	25	30	35	40	45
# 6	6	12	18	24	30	36	42	48	54
# 7	7	14	21	28	35	42	49	56	63
# 8	8	16	24	32	40	48	56	64	72
# 9	9	18	27	36	45	54	63	72	81
# 10	10	20	30	40	50	60	70	80	90

Summary 4 Chapter 2 Lesson 3
Multiplication numbers from 10 to 19

Multiply numbers in row

$5 \times 10 = 50$ $10 \times 5 = 50$ $5 \times 11 = 55$ $11 \times 5 = 55$

1) 15×9
2) $15 = 10 + 5$
3) $10 \times 9 = 90$
4) $5 \times 9 = 45$
5) $90 + 45 = 135$
6) $15 \times 9 = 135$

Multiply numbers

	a	b	c	d	e
# 1	10×7	11×7	12×7	13×7	14×7
# 2	10×8	11×8	12×8	13×8	14×8
# 3	10×9	11×9	12×9	13×9	14×9
# 4	15×2	16×2	17×2	18×2	19×2
# 5	15×3	16×3	17×3	18×3	19×3

Multiply numbers in column

$$\begin{array}{r} 15 \\ \times 9 \\ \hline 45 \\ +90 \\ \hline 135 \end{array}$$

1) $9 \times 5 = 45$
2) $9 \times 10 = 90$

Multiply numbers

	a	b	c	d	e
# 6	15×7	16×7	17×7	18×7	19×7
# 7	15×8	16×8	17×8	18×8	19×8
# 8	15×9	16×9	17×9	18×9	19×9

Summary 4 Chapter 2 Lesson 4
Multiplication numbers from 20 to 29
Multiply numbers in row

1) 24×9
2) $24 = 20 + 4$
3) $20 \times 9 = 180$
4) $4 \times 9 = 36$
5) $180 + 36 = 216$
6) $24\times 9 = 216$

Multiply numbers

	a	b	c	d	e
# 1	20×7	21×7	22×7	23×7	24×7
# 2	20×8	21×8	22×8	23×8	24×8
# 3	20×9	21×9	22×9	23×9	24×9
# 4	25×2	26×2	27×2	28×2	29×2
# 5	25×3	26×3	27×3	28×3	29×3

Multiply numbers in column

$$\begin{array}{r} {}^{6}\\ 27 \\ \underline{\times 9} \\ 243 \end{array}$$

1) $9 \times 7 = 63$
2) Put 3 below 9
3) Put 6 above 2
4) $9 \times 2 = 18$
5) $18 + 6 = 24$
6) Put 24 ahead 3

Multiply numbers

	a	b	c	d	
# 6	$\begin{array}{r} 25 \\ \times \underline{7} \end{array}$	$\begin{array}{r} 26 \\ \times \underline{7} \end{array}$	$\begin{array}{r} 27 \\ \times \underline{7} \end{array}$	$\begin{array}{r} 28 \\ \times \underline{7} \end{array}$	$\begin{array}{r} 29 \\ \times \underline{7} \end{array}$
# 7	$\begin{array}{r} 25 \\ \times \underline{8} \end{array}$	$\begin{array}{r} 26 \\ \times \underline{8} \end{array}$	$\begin{array}{r} 27 \\ \times \underline{8} \end{array}$	$\begin{array}{r} 28 \\ \times \underline{8} \end{array}$	$\begin{array}{r} 29 \\ \times \underline{8} \end{array}$
# 8	$\begin{array}{r} 25 \\ \times \underline{9} \end{array}$	$\begin{array}{r} 26 \\ \times \underline{9} \end{array}$	$\begin{array}{r} 27 \\ \times \underline{9} \end{array}$	$\begin{array}{r} 28 \\ \times \underline{9} \end{array}$	$\begin{array}{r} 29 \\ \times \underline{9} \end{array}$

Summary 4 Chapter 2 Lesson 5
Multiplication numbers from 30 to 39
Multiply numbers in row

1) 35×9
2) $35 = 30 + 5$
3) $30 \times 9 = 270$
4) $5 \times 9 = 45$
5) $270 + 45 = 315$
6) $35\times 9 = 315$

Multiply numbers

	a	b	c	d	e
# 1	30×7	31×7	32×7	33×7	34×7
# 2	30×8	31×8	32×8	33×8	34×8
# 3	30×9	31×9	32×9	33×9	34×9
# 4	35×2	36×2	37×2	38×2	39×2
# 5	35×3	36×3	37×3	38×3	39×3

Multiply numbers in column

$$\begin{array}{r} {\scriptstyle 7} \\ 38 \\ \times\underline{9} \\ 342 \end{array}$$

1) $9 \times 8 = 72$
2) Put 2 below 9
3) Put 7 above 3
4) $9 \times 3 = 27$
5) $27 + 7 = 34$
6) Put 34 ahead 2

Multiply numbers

	a	b	c	d	e
# 6	$35 \times \underline{7}$	$36 \times \underline{7}$	$37 \times \underline{7}$	$38 \times \underline{7}$	$39 \times \underline{7}$
# 7	$35 \times \underline{8}$	$36 \times \underline{8}$	$37 \times \underline{8}$	$38 \times \underline{8}$	$39 \times \underline{8}$
# 8	$35 \times \underline{9}$	$36 \times \underline{9}$	$37 \times \underline{9}$	$38 \times \underline{9}$	$39 \times \underline{9}$

Summary 4 Chapter 2 Lesson 6
Multiplication numbers from 40 to 49
Multiply numbers in row

1) 43×9
2) $43 = 40 + 3$
3) $40 \times 9 = 360$
4) $3 \times 9 = 27$
5) $360 + 27 = 387$
6) $43\times 9 = 387$

Multiply numbers

	a	b	c	d	e
# 1	40×7	41×7	42×7	43×7	44×7
# 2	40×8	41×8	42×8	43×8	44×8
# 3	40×9	41×9	42×9	43×9	44×9
# 4	45×2	46×2	47×2	48×2	49×2
# 5	45×3	46×3	47×3	48×3	49×3

Multiply numbers in column

$$\begin{array}{r} {}^{2} \\ 43 \\ \times\underline{9} \\ 387 \end{array}$$

1) $9 \times 3 = 27$
2) Put 7 below 9
3) Put 2 above 4
4) $9 \times 4 = 36$
5) $36 + 2 = 38$
6) Put 38 ahead 7

Multiply numbers

	a	b	c	d	e
# 6	$45 \times \underline{7}$	$46 \times \underline{7}$	$47 \times \underline{7}$	$48 \times \underline{7}$	$49 \times \underline{7}$
# 7	$45 \times \underline{8}$	$46 \times \underline{8}$	$47 \times \underline{8}$	$48 \times \underline{8}$	$49 \times \underline{8}$
# 8	$45 \times \underline{9}$	$46 \times \underline{9}$	$47 \times \underline{9}$	$48 \times \underline{9}$	$49 \times \underline{9}$

Summary 4 Chapter 2 Lesson 7
Multiplication from 50 to 59
Multiply numbers in row

1) 57×9
2) $57 = 50 + 7$
3) $50 \times 9 = 450$
4) $7 \times 9 = 63$
5) $450 + 63 = 513$
6) $57 \times 9 = 513$

Multiply numbers

	a	b	c	d	e
# 1	50×7	51×7	52×7	53×7	54×7
# 2	50×8	51×8	52×8	53×8	54×8
# 3	50×9	51×9	52×9	53×9	54×9
# 4	55×2	56×2	57×2	58×2	59×2
# 5	55×3	56×3	57×3	58×3	59×3

Multiply numbers in column

$$\begin{array}{r} {}^{6}\\ 57 \\ \underline{\times 9} \\ 513 \end{array}$$

1) $9 \times 7 = 63$
2) Put 3 below 9
3) Put 6 above 5
4) $9 \times 5 = 45$
5) $45 + 6 = 51$
6) Put 51 ahead 3

Multiply numbers

	a	b	c	d	e
# 6	$55 \times \underline{7}$	$56 \times \underline{7}$	$57 \times \underline{7}$	$58 \times \underline{7}$	$59 \times \underline{7}$
# 7	$55 \times \underline{8}$	$56 \times \underline{8}$	$57 \times \underline{8}$	$58 \times \underline{8}$	$59 \times \underline{8}$
# 8	$55 \times \underline{9}$	$56 \times \underline{9}$	$57 \times \underline{9}$	$58 \times \underline{9}$	$59 \times \underline{9}$

Summary 4 Chapter 2 Lesson 8
Multiplication from 60 to 69
Multiply numbers in row

1) 63×9
2) $63 = 60 + 3$
3) $60 \times 9 = 540$
4) $3 \times 9 = 27$
5) $540 + 27 = 567$
6) $63 \times 9 = 567$

Multiply numbers

	a	b	c	d	e
# 1	60×7	61×7	62×7	63×7	64×7
# 2	60×8	61×8	62×8	63×8	64×8
# 3	60×9	61×9	62×9	63×9	64×9
# 4	65×2	66×2	67×2	68×2	69×2
# 5	65×3	66×3	67×3	68×3	69×3

Multiply numbers in column

$$\begin{array}{r} {}^{2}\\ 63 \\ \times\underline{9} \\ 567 \end{array}$$

1) $9 \times 3 = 27$
2) Put 7 below 9
3) Put 2 above 6
4) $9 \times 6 = 54$
5) $54 + 2 = 56$
6) Put 56 ahead 7

Multiply numbers

	a	b	c	d	e
# 6	$65 \times \underline{7}$	$66 \times \underline{7}$	$67 \times \underline{7}$	$68 \times \underline{7}$	$69 \times \underline{7}$
# 7	$65 \times \underline{8}$	$66 \times \underline{8}$	$67 \times \underline{8}$	$68 \times \underline{8}$	$69 \times \underline{8}$
# 8	$65 \times \underline{9}$	$66 \times \underline{9}$	$67 \times \underline{9}$	$68 \times \underline{9}$	$69 \times \underline{9}$

Summary 4 Chapter 2 Lesson 9
Multiplication numbers from 70 – 79
Multiply numbers in row

1) 73×9
2) $73 = 70 + 3$
3) $70 \times 9 = 630$
4) $3 \times 9 = 27$
5) $630 + 27 = 657$
6) $73\times 9 = 657$

Multiply numbers

	a	b	c	d	e
# 1	70×7	71×7	72×7	73×7	74×7
# 2	70×8	71×8	72×8	73×8	74×8
# 3	70×9	71×9	72×9	73×9	74×9
# 4	75×2	76×2	77×2	78×2	79×2
# 5	75×3	76×3	77×3	78×3	79×3

Multiply numbers in column

$$\begin{array}{r} {}^{2} \\ 73 \\ \underline{\times 9} \\ 657 \end{array}$$

1) $9 \times 3 = 27$
2) Put 7 below 9
3) Put 2 above 7
4) $9 \times 7 = 63$
5) $63 + 2 = 65$
6) Put 65 ahead 7

Multiply numbers

	a	b	c	d	e
# 6	$75 \times \underline{7}$	$76 \times \underline{7}$	$77 \times \underline{7}$	$78 \times \underline{7}$	$79 \times \underline{7}$
# 7	$75 \times \underline{8}$	$76 \times \underline{8}$	$77 \times \underline{8}$	$78 \times \underline{8}$	$79 \times \underline{8}$
# 8	$75 \times \underline{9}$	$76 \times \underline{9}$	$77 \times \underline{9}$	$78 \times \underline{9}$	$79 \times \underline{9}$

Summary 4 Chapter 2 Lesson 10
Multiplication numbers from 80 to 89
Multiply numbers in row

1) 87× 9
2) 87 = 80 + 7
3) 80 × 9 = 720
4) 7 × 9 = 63
5) 720 + 63 = 783
6) 87× 9 = 783

Multiply numbers

	a	b	c	d	e
# 1	80×7	81×7	82×7	83×7	84×7
# 2	80×8	81×8	82×8	83×8	84×8
# 3	80×9	81×9	82×9	83×9	84×9
# 4	85×2	86×2	87×2	88×2	89×2
# 5	85×3	86×3	87×3	88×3	89×3

Multiply numbers in column

```
 6
 87
×9
---
783
```

1) 9 × 7 = 63
2) Put 3 below 9
3) Put 6 above 8
4) 9 × 8 = 72
5) 72 + 6 = 78
6) Put 78 ahead 3

Multiply numbers

	a	b	c	d	e
# 6	85 ×7	86 ×7	87 ×7	88 ×7	89 ×7
# 7	85 ×8	86 ×8	87 ×8	88 ×8	89 ×8
# 8	85 ×9	86 ×9	87 ×9	88 ×9	89 ×9

Summary 4 Chapter 2 Lesson 11
Multiplication from 90 to 99
Multiply numbers in row

1) 96× 7
2) 96 = 90 + 6
3) 90 × 7 = 630
4) 6 × 7 = 42
5) 630 + 42 = 672
6) 96× 7 = 672

Multiply numbers

	a	b	c	d	e
# 1	90×7	91×7	92×7	93×7	94×7
# 2	90×8	91×8	92×8	93×8	94×8
# 3	90×9	91×9	92×9	93×9	94×9
# 4	95×2	96×2	97×2	98×2	99×2
# 5	95×3	96×3	97×3	98×3	99×3

Multiply numbers in column

```
  4
 96
×7
---
672
```

1) 7 × 6 = 42
2) Put 2 below 7
3) Put 4 above 9
4) 7 × 9 = 63
5) 63 + 4 = 67
6) Put 67 ahead 2

Multiply numbers

	a	b	c	d	e
# 6	95 ×7	96 ×7	97 ×7	98 ×7	99 ×7
# 7	95 ×8	96 ×8	97 ×8	98 ×8	99 ×8
# 8	95 ×9	96 ×9	97 ×9	98 ×9	99 ×9

Summary 4 Chapter 2 Lesson 12

Multiplication and Addition numbers

To multiply and add numbers always first multiply than add

$$9 \times 6 + 9 = 54 + 9 = 63$$

Evaluate. Review 1 A

	a	b	c	d
# 1	$2 \times 6 + 5$	$3 \times 6 + 5$	$4 \times 6 + 5$	$5 \times 6 + 5$
# 2	$2 \times 7 + 6$	$3 \times 7 + 6$	$4 \times 7 + 6$	$5 \times 7 + 6$
# 3	$2 \times 8 + 7$	$3 \times 8 + 7$	$4 \times 8 + 7$	$5 \times 8 + 7$
# 4	$2 \times 9 + 8$	$3 \times 9 + 8$	$4 \times 9 + 8$	$5 \times 9 + 8$
# 5	$2 \times 10 + 9$	$3 \times 10 + 9$	$4 \times 10 + 9$	$5 \times 10 + 9$

Multiply and add numbers

$$9 \overset{2}{+} 9 \overset{1}{\times} 9 = 9 + 81 = 90$$

	a	b	c	d
# 6	$5 + 6 \times 6$	$5 + 7 \times 6$	$5 + 8 \times 6$	$5 + 9 \times 6$
# 7	$6 + 6 \times 7$	$6 + 7 \times 7$	$6 + 8 \times 7$	$6 + 9 \times 7$
# 8	$7 + 6 \times 8$	$7 + 7 \times 8$	$7 + 8 \times 8$	$7 + 9 \times 8$
# 9	$8 + 6 \times 9$	$8 + 7 \times 9$	$8 + 8 \times 9$	$8 + 9 \times 9$
# 10	$9 + 6 \times 10$	$9 + 7 \times 10$	$9 + 8 \times 10$	$9 + 9 \times 10$

Multiplication and Subtraction numbers

To multiply and subtraction numbers always first multiply than subtract

$$5 \times 10 - 9 = 50 - 9 = 41$$

Evaluate. Review 1 B

	a	b	c	d
# 1	$2 \times 6 - 5$	$3 \times 6 - 5$	$4 \times 6 - 5$	$5 \times 6 - 5$
# 2	$2 \times 7 - 6$	$3 \times 7 - 6$	$4 \times 7 - 6$	$5 \times 7 - 6$
# 3	$2 \times 8 - 7$	$3 \times 8 - 7$	$4 \times 8 - 7$	$5 \times 8 - 7$
# 4	$2 \times 9 - 8$	$3 \times 9 - 8$	$4 \times 9 - 8$	$5 \times 9 - 8$
# 5	$2 \times 10 - 9$	$3 \times 10 - 9$	$4 \times 10 - 9$	$5 \times 10 - 9$

Multiply and Subtract numbers

$$100 \overset{2}{-} 9 \overset{1}{\times} 10 = 100 - 90 = 10$$

	a	b	c	d
# 6	$10 - 6 \times 1$	$10 - 7 \times 1$	$10 - 8 \times 1$	$10 - 9 \times 1$
# 7	$20 - 6 \times 2$	$20 - 7 \times 2$	$20 - 8 \times 2$	$20 - 9 \times 2$
# 8	$30 - 6 \times 3$	$30 - 7 \times 3$	$30 - 8 \times 3$	$30 - 9 \times 3$
# 9	$40 - 6 \times 4$	$40 - 7 \times 4$	$40 - 8 \times 4$	$40 - 9 \times 4$
# 10	$50 - 6 \times 5$	$50 - 7 \times 5$	$50 - 8 \times 5$	$50 - 9 \times 5$

Summary 4 Chapter 2 Lesson 13

Summary 1

Three-digit numbers

Three-digit numbers content three digits.
This is numbers 100, 101, … 999
Least three-digit number: 100.
Greatest three-digit number: 999
Digits for three-digit numbers from right to the left:
Ones, tens, and hundreds 468 = 400 + 60 + 8

One-digit numbers less than two or three-digit numbers
Three-digit numbers greater than one or two-digit numbers

To compare numbers, compare first hundred, after that tens, then ones   423 < 456   789 > 786
Two numbers we can add in any order. The sum will be the same.

200 + 300 = 500   300+200=500

The sum of a number and zero is the same number

500+0 = 500   0+500=500

Difference any number and zero is the same number:
500 – 0 = 500

Difference two the same number is zero: 500 – 500 = 0

Adding three-digit numbers: Add the ones. Add the tens. Add the hundreds.

Subtracting three-digit numbers:
Subtract the ones. Subtract the tens. Subtract the hundreds.

		11	71414
246	678	154	85 4
+ 351	- 432	+789	-28 6
597	246	943	56 8

Summary 4 Chapter 2 Lesson 13
Summary 2
Multiplication numbers

Third operation with numbers is multiplication × or ·
Multiplication is the operation of repeated addition
$2\times 3 = 2 + 2 + 2 = 6$ or $2 \cdot 3 = 2 + 2 + 2 = 6$

First number in multiplication called the factor	**2**
Second number in multiplication called the factor	**3**
Result of multiplication called the product	**6**

Properties of multiplication
Two numbers we can multiply in any order.
The product will the same.
$2 \times 3 = 2 + 2 + 2 = 6$ $3 \times 2 = 3 + 3 = 6$
The product of a number and zero is zero
$5 \times 0 = 0$ $0 \times 5 = 0$
The product of a number and one is the same number
$5 \times 1 = 5$ $1 \times 5 = 5$

Division two-digit and one-digit numbers
Fourth operation with numbers is division ÷ or :
$6 \div 2 = 3$ $6 : 3 = 2$
Division means separated number by some equal groups

First number in division is called dividend	**6**
Second number in division is called divisor	**2**
Result of division is called quotient	**3**

Properties of division

Any number divided by one is the same number		**$5 \div 1 = 5$**
Any number divided by the itself is one		**$5 \div 5 = 1$**
Zero divided by any number is zero		**$0 \div 5 = 0$**
Division by zero is not allowed	**$5 \div 0$**	**No number**

Summary 4 Chapter 2 Lesson 13
Review 2 A
Multiply the numbers in column

# 1 $\begin{array}{r} 16 \\ \times 9 \\ \hline \end{array}$	# 2 $\begin{array}{r} 18 \\ \times 7 \\ \hline \end{array}$	# 3 $\begin{array}{r} 26 \\ \times 7 \\ \hline \end{array}$	# 4 $\begin{array}{r} 28 \\ \times 6 \\ \hline \end{array}$
# 5 $\begin{array}{r} 36 \\ \times 7 \\ \hline \end{array}$	# 6 $\begin{array}{r} 38 \\ \times 9 \\ \hline \end{array}$	# 7 $\begin{array}{r} 46 \\ \times 8 \\ \hline \end{array}$	# 8 $\begin{array}{r} 48 \\ \times 7 \\ \hline \end{array}$
# 9 $\begin{array}{r} 56 \\ \times 7 \\ \hline \end{array}$	# 10 $\begin{array}{r} 58 \\ \times 5 \\ \hline \end{array}$	# 11 $\begin{array}{r} 65 \\ \times 9 \\ \hline \end{array}$	# 12 $\begin{array}{r} 68 \\ \times 9 \\ \hline \end{array}$
# 13 $\begin{array}{r} 76 \\ \times 8 \\ \hline \end{array}$	# 14 $\begin{array}{r} 78 \\ \times 9 \\ \hline \end{array}$	# 15 $\begin{array}{r} 86 \\ \times 7 \\ \hline \end{array}$	# 16 $\begin{array}{r} 89 \\ \times 7 \\ \hline \end{array}$
# 17 $\begin{array}{r} 92 \\ \times 6 \\ \hline \end{array}$	# 18 $\begin{array}{r} 95 \\ \times 7 \\ \hline \end{array}$	# 19 $\begin{array}{r} 96 \\ \times 7 \\ \hline \end{array}$	# 20 $\begin{array}{r} 98 \\ \times 6 \\ \hline \end{array}$

Summary 4 Chapter 2 Lesson 13

Review 2 B

Multiply the numbers in column

# 1	17 × 9	# 2	18 × 6	# 3	26 × 9	# 4	28 × 7
# 5	36 × 5	# 6	38 × 7	# 7	46 × 8	# 8	48 × 7
# 9	56 × 8	# 10	58 × 6	# 11	67 × 9	# 12	68 × 7
# 13	76 × 9	# 14	78 × 6	# 15	86 × 5	# 16	87 × 9
# 17	92 × 8	# 18	94 × 7	# 19	96 × 8	# 20	98 × 5

Summary 4 Chapter 2 Lesson 14
Multiplication numbers from 10 to 99

Test 2 A

Multiply the numbers in column

# 1 $\begin{array}{r} 15 \\ \underline{\times 9} \end{array}$	# 2 $\begin{array}{r} 16 \\ \underline{\times 7} \end{array}$	# 3 $\begin{array}{r} 25 \\ \underline{\times 7} \end{array}$	# 4 $\begin{array}{r} 27 \\ \underline{\times 6} \end{array}$
# 5 $\begin{array}{r} 34 \\ \underline{\times 7} \end{array}$	# 6 $\begin{array}{r} 36 \\ \underline{\times 9} \end{array}$	# 7 $\begin{array}{r} 45 \\ \underline{\times 8} \end{array}$	# 8 $\begin{array}{r} 46 \\ \underline{\times 7} \end{array}$
# 9 $\begin{array}{r} 54 \\ \underline{\times 7} \end{array}$	# 10 $\begin{array}{r} 56 \\ \underline{\times 5} \end{array}$	# 11 $\begin{array}{r} 64 \\ \underline{\times 9} \end{array}$	# 12 $\begin{array}{r} 66 \\ \underline{\times 9} \end{array}$
# 13 $\begin{array}{r} 74 \\ \underline{\times 8} \end{array}$	# 14 $\begin{array}{r} 76 \\ \underline{\times 9} \end{array}$	# 15 $\begin{array}{r} 84 \\ \underline{\times 7} \end{array}$	# 16 $\begin{array}{r} 85 \\ \underline{\times 7} \end{array}$
# 17 $\begin{array}{r} 94 \\ \underline{\times 6} \end{array}$	# 18 $\begin{array}{r} 96 \\ \underline{\times 7} \end{array}$	# 19 $\begin{array}{r} 94 \\ \underline{\times 7} \end{array}$	# 20 $\begin{array}{r} 96 \\ \underline{\times 6} \end{array}$

Summary 4 Chapter 2 Lesson 14
Multiplication numbers from 10 to 99

Test 2 B

Multiply the numbers in column

# 1 $\begin{array}{r} 14 \\ \underline{\times 9} \end{array}$	# 2 $\begin{array}{r} 19 \\ \underline{\times 6} \end{array}$	# 3 $\begin{array}{r} 25 \\ \underline{\times 9} \end{array}$	# 4 $\begin{array}{r} 26 \\ \underline{\times 7} \end{array}$
# 5 $\begin{array}{r} 34 \\ \underline{\times 5} \end{array}$	# 6 $\begin{array}{r} 36 \\ \underline{\times 7} \end{array}$	# 7 $\begin{array}{r} 48 \\ \underline{\times 8} \end{array}$	# 8 $\begin{array}{r} 46 \\ \underline{\times 7} \end{array}$
# 9 $\begin{array}{r} 54 \\ \underline{\times 8} \end{array}$	# 10 $\begin{array}{r} 56 \\ \underline{\times 6} \end{array}$	# 11 $\begin{array}{r} 65 \\ \underline{\times 9} \end{array}$	# 12 $\begin{array}{r} 69 \\ \underline{\times 7} \end{array}$
# 13 $\begin{array}{r} 74 \\ \underline{\times 9} \end{array}$	# 14 $\begin{array}{r} 76 \\ \underline{\times 6} \end{array}$	# 15 $\begin{array}{r} 84 \\ \underline{\times 5} \end{array}$	# 16 $\begin{array}{r} 86 \\ \underline{\times 9} \end{array}$
# 17 $\begin{array}{r} 94 \\ \underline{\times 8} \end{array}$	# 18 $\begin{array}{r} 96 \\ \underline{\times 7} \end{array}$	# 19 $\begin{array}{r} 94 \\ \underline{\times 8} \end{array}$	# 20 $\begin{array}{r} 96 \\ \underline{\times 5} \end{array}$

Summary 4 Answers Chapter 2

Lesson 2

	a, b, c, d		a, b, c, d
# 1	2, 3, 4, 5	# 11	6, 7, 8, 9
# 2	4, 6, 8, 10	# 12	12, 14, 16, 18
# 3	6, 9, 12, 15	# 13	18, 21, 24, 27
# 4	8, 12, 16, 20	# 14	24, 28, 32, 36
# 5	10, 15, 20, 25	# 15	30, 35, 40, 45
# 6	12, 18, 24, 30	# 16	36, 42, 48, 54
# 7	14, 21, 28, 35	# 17	42, 49, 56, 63
# 8	16, 24, 32, 40	# 18	48, 56, 64, 72
# 9	18, 27, 36, 45	# 19	54, 63, 72, 81
# 10	20, 30, 40, 50	# 20	60, 70, 80, 90

Lesson 3

	a, b, c, d, e		a, b, c, d, e
# 1	70, 77, 84, 91, 98	# 5	45, 48, 51, 54, 57
# 2	80, 88, 96, 104, 112	# 6	105, 112, 119, 126, 133
# 3	90, 99, 108, 117, 126	# 7	120, 128, 136, 144, 152
# 4	30, 32, 34, 36, 38	# 8	135, 144, 153, 162, 171

Lesson 4

	a, b, c, d, e		a, b, c, d, e
# 1	140, 147, 154, 161, 168	# 5	75, 78, 81, 84, 87
# 2	160, 168, 176, 184, 192	# 6	175, 182, 189, 196, 203
# 3	180, 189, 198, 207, 216	# 7	200, 208, 216, 224, 232
# 4	50, 52, 54, 56, 58	# 8	225, 234, 243, 252, 261

Lesson 5

	a, b, c, d, e		a, b, c, d, e
# 1	210, 217, 224, 231, 238	# 5	105, 108, 111, 114, 117
# 2	240, 248, 256, 264, 272	# 6	245, 252, 259, 266, 273
# 3	270, 279, 288, 297, 306	# 7	280, 288, 296, 304, 312
# 4	70, 72, 74, 76, 78	# 8	315, 324, 333, 342, 351

Summary 4 Answers Chapter 2

Lesson 6

	a, b, c, d, e		a, b, c, d, e
# 1	280, 287, 294, 301, 308	# 5	135, 138, 141, 144, 147
# 2	320, 328, 336, 344, 352	# 6	315, 322, 329, 336, 343
# 3	360, 369, 378, 387, 396	# 7	360, 368, 376, 384, 392
# 4	90, 92, 94, 96, 98	# 8	405, 414, 423, 432, 441

Lesson 7

	a, b, c, d, e		a, b, c, d, e
# 1	350, 357, 364, 371, 378	# 5	165, 168, 171, 174, 177
# 2	400, 408, 416, 424, 432	# 6	385, 392, 399, 406, 413
# 3	450, 459, 468, 477, 486	# 7	440, 448, 456, 464, 472
# 4	110, 112, 114, 116, 118	# 8	495, 504, 513, 522, 531

Lesson 8

	a, b, c, d, e		a, b, c, d, e
# 1	420, 427, 434, 441, 448	# 5	195, 198, 201, 204, 207
# 2	480, 488, 496, 504, 512	# 6	455, 462, 469, 476, 483
# 3	540, 549, 558, 567, 576	# 7	520, 528, 536, 544, 552
# 4	130, 132, 134, 136, 138	# 8	585, 594, 603, 612, 621

Lesson 9

	a, b, c, d, e		a, b, c, d, e
# 1	490, 497, 504, 511, 518	# 5	225, 228, 231, 234, 237
# 2	560, 568, 576, 584, 592	# 6	525, 532, 539, 546, 553
# 3	630, 639, 648, 657, 666	# 7	600, 608, 616, 624, 632
# 4	150, 152, 154, 156, 158	# 8	675, 684, 693, 702, 711

Summary 4 Answers Chapter 2

Lesson 10

	a, b, c, d, e		a, b, c, d, e
# 1	560, 567, 574, 581, 588	# 5	255, 258, 261, 264, 267
# 2	640, 648, 656, 664, 672	# 6	595, 602, 609, 616, 623
# 3	720, 729, 738, 747, 756	# 7	680, 688, 696, 704, 712
# 4	170, 172, 174, 176, 178	# 8	765, 774, 783, 792, 801

Lesson 11

	a, b, c, d, e		a, b, c, d, e
# 1	630, 637, 644, 651, 658	# 5	285, 288, 291, 294, 297
# 2	720, 728, 736, 744, 752	# 6	665, 672, 679, 686, 693
# 3	810, 819, 828, 837, 846	# 7	760, 768, 776, 784, 792
# 4	190, 192, 194, 196, 198	# 8	855, 864, 873, 882, 891

Lesson 12

Review 1 A

	a	b	c	d
# 1	17	23	29	35
# 2	20	27	34	41
# 3	23	31	39	47
# 4	26	35	44	53
# 5	29	39	49	59
# 6	41	47	53	59
# 7	48	55	62	69
# 8	55	63	71	79
# 9	62	71	80	89
# 10	69	79	89	99

Review 1 B

	a	b	c	d
# 1	7	13	19	25
# 2	8	15	22	29
# 3	9	17	25	33
# 4	10	19	28	37
# 5	11	21	31	41
# 6	4	3	2	1
# 7	8	6	4	2
# 8	12	9	6	3
# 9	16	12	8	4
# 10	20	15	10	5

Summary 4 Answers Chapter 2

Lesson 13 Review 2 A

# 1	144	# 6	342	# 11	585	# 16	623
# 2	126	# 7	368	# 12	612	# 17	552
# 3	182	# 8	336	# 13	608	# 18	665
# 4	168	# 9	392	# 14	702	# 19	672
# 5	252	# 10	290	# 15	602	# 20	588

Lesson 13 Review 2 B

# 1	153	# 6	266	# 11	603	# 16	783
# 2	108	# 7	368	# 12	476	# 17	736
# 3	234	# 8	336	# 13	684	# 18	658
# 4	196	# 9	448	# 14	468	# 19	768
# 5	180	# 10	348	# 15	430	# 20	490

Lesson 14 Test 2 A

# 1	135	# 2	112	# 3	175	# 4	162	# 5	238
# 6	324	# 7	360	# 8	322	# 9	378	# 10	280
# 11	576	# 12	594	# 13	592	# 14	684	# 15	588
# 16	595	# 17	564	# 18	672	# 19	658	# 20	576

Test 2 B

# 1	126	# 2	114	# 3	225	# 4	182	# 5	170
# 6	252	# 7	384	# 8	322	# 9	432	# 10	336
# 11	585	# 12	483	# 13	666	# 14	456	# 15	420
# 16	774	# 17	752	# 18	672	# 19	752	# 20	480

Summary 4

Chapter 3

Division

Numbers 1

Summary 4 Chapter 3 Contents
Division Numbers. Lesson 1 – 20
Basic Problems

Work with lesson 1, read rules and examples, 2) Solve the problems from contents, 3) Check answers in the end of chapter 1, 4) Work with the next lessons 2, 3, 4, 5, ..., 5) Solve Test 1 A, 6) Grading Scale: A =10 points, B=9points, C=8 points, D=7 points, F=0-6 points, 7) If you solve 10 problems correctly, you get an A, 8) If have an A, you can move on to next chapter, 9) If your grade is lower than an A, work with the mistakes, then solve Test 1B, 10) When you finish all of the chapters, solve Final Exam in last chapter. Lesson = L

Answers Chapter 3
Summary 4 Chapter 3 Lesson 1
Division numbers 1

Fourth operation with numbers is division ÷ or :
Division means repeated subtraction. Result of division means how many times we need to subtract the same number to get zero

$6 - 2 - 2 - 2 = 0$ $6 \div 2 = 3$ $6 - 3 - 3 = 0$ $6 : 3 = 2$

1 2 3 1 2

$2 \times 3 = 6$ $6 \div 2 = 3$ $6 \div 3 = 2$

Table of multiplication

Memorize the Table of Multiplication. Number 5 from the left column multiply by number 4 from first row to get, answer 20

	×	2	3	<u>4</u>	5	6	7	8	9
# 1	1	2	3	4	5	6	7	8	9
# 2	2	4	6	8	10	12	14	16	18
# 3	3	6	9	12	15	18	21	24	27
# 4	4	8	12	16	20	24	28	32	36
# 5	<u>5</u>	10	15	<u>20</u>	25	30	35	40	45
# 6	6	12	18	24	30	36	42	48	54
# 7	7	14	21	28	35	42	49	56	63
# 8	8	16	24	32	40	48	56	64	72
# 9	9	18	27	36	45	54	63	72	81
# 10	10	20	30	40	50	60	70	80	90

Divide the numbers using table multiplication

	a	b	c	d
# 1	2 ÷ 2	12 ÷ 2	4 ÷ 4	24 ÷ 4
# 2	4 ÷ 2	14 ÷ 2	8 ÷ 4	28 ÷ 4
# 3	6 ÷ 2	16 ÷ 2	12 ÷ 4	32 ÷ 4
# 4	8 ÷ 2	18 ÷ 2	16 ÷ 4	36 ÷ 4
# 5	10÷2	20 ÷ 2	20 ÷ 4	40 ÷ 4
# 6	3 ÷ 3	18 ÷ 3	5 ÷ 5	30 ÷ 5
# 7	6 ÷ 3	21 ÷ 3	10 ÷ 5	35 ÷ 5
# 8	9 ÷ 3	24 ÷ 3	15 ÷ 5	40 ÷ 5
# 9	12÷3	27 ÷ 3	20 ÷ 5	45 ÷ 5
# 10	15÷3	30 ÷ 3	25 ÷ 5	50 ÷ 5

Summary 4 Chapter 3 Lesson 2
Properties of division 2

Division also means separated number by some equal groups

6÷2=3 **2×3 = 6**

First number in division called the dividend 6

Second number in division called the divisor 2

Result of division called the quotient 3

Any number divided by one is the same number 5 ÷ 1 = 5

Any number divided by the itself is one 5 ÷ 5 = 1

Zero divided by any number is zero 0 ÷ 5 = 0

Division by zero is not allowed 5 ÷ 0 No number

To check multiplication use division

2×3=6 Check 6÷2=3 or 6÷3=2

To check division use multiplication and division

6÷2=3 Check 2×3=6 or 6÷3=2

Division without remainder R = 0 **Division with remainder R > 0**

Dividend 35, Divisor 7, Quotient 5, **Dividend 36, Divisor 7, Quotient 5**

```
  5
7)35
 -35      R = 0
   0
```

```
  5
7)36
 -35      R = 1
   1
```

Complete the Table of Division by using Table of Multiplication

	a	b	c	d	e	f	g	h	i
	×	2		4		6		8	
#1	1								9
#2									18
#3	3							24	
#4								32	
#5	5						35		
#6						36			
#7	7				35				
#8				32					
#9	9		27						
#10		20							

Summary 4 Chapter 3 Lesson 3
Division number with quotient from 10 to 19
Divide numbers in column 3
Divide numbers

	a	b	c	d
# 1	20 ÷ 2	33 ÷ 3	76 ÷ 4	90 ÷ 5
# 2	24 ÷ 2	39 ÷ 3	68 ÷ 4	80 ÷ 5
# 3	28 ÷ 2	45 ÷ 3	60 ÷ 4	70 ÷ 5
# 4	32 ÷ 2	51 ÷ 3	52 ÷ 4	60 ÷ 5
# 5	36 ÷ 2	57 ÷ 3	44 ÷ 4	50 ÷ 5
# 6	22 ÷ 2	30 ÷ 3	72 ÷ 4	95 ÷ 5
# 7	26 ÷ 2	36 ÷ 3	64 ÷ 4	85 ÷ 5
# 8	30 ÷ 2	42 ÷ 3	56 ÷ 4	75 ÷ 5
# 9	34 ÷ 2	48 ÷ 3	48 ÷ 4	65 ÷ 5
# 10	38 ÷ 2	54 ÷ 3	40 ÷ 4	55 ÷ 5

Divide numbers in column

$9 \times 1 = 9 \qquad 9 \times 2 = 18$

$$\begin{array}{r} 19 \\ 9\,\overline{)\,171} \\ \underline{-9} \\ 81 \\ \underline{-81} \\ 0 \end{array}$$

1) $9 \times 1 = 9$
2) $17 - 9 = 8$
3) Bring down the 1
4) $9 \times 9 = 81$
5) $81 - 81 = 0$
6) R = 0

Check

$$\begin{array}{r} 8 \\ 19 \\ \underline{\times 9} \\ 171 \end{array}$$

Divide numbers

	a	b	c	d
# 11	60 ÷ 6	77 ÷ 7	152 ÷ 8	162 ÷ 9
# 12	72 ÷ 6	91 ÷ 7	136 ÷ 8	144 ÷ 9
# 13	84 ÷ 6	105 ÷ 7	120 ÷ 8	126 ÷ 9
# 14	96 ÷ 6	119 ÷ 7	104 ÷ 8	108 ÷ 9
# 15	108 ÷ 6	133 ÷ 7	88 ÷ 8	90 ÷ 9
# 16	66 ÷ 6	70 ÷ 7	144 ÷ 8	171 ÷ 9
# 17	78 ÷ 6	84 ÷ 7	128 ÷ 8	153 ÷ 9
# 18	90 ÷ 6	98 ÷ 7	112 ÷ 8	135 ÷ 9
# 19	102 ÷ 6	112 ÷ 7	96 ÷ 8	117 ÷ 9
# 20	114 ÷ 6	126 ÷ 7	80 ÷ 8	99 ÷ 9

Summary 4 Chapter 3 Lesson 4

Division numbers with quotient from 20 to 29

Divide numbers in column 4

Divide numbers

	a	b	c	d
# 1	$40 \div 2$	$63 \div 3$	$116 \div 4$	$140 \div 5$
# 2	$44 \div 2$	$69 \div 3$	$108 \div 4$	$130 \div 5$
# 3	$48 \div 2$	$75 \div 3$	$100 \div 4$	$120 \div 5$
# 4	$52 \div 2$	$81 \div 3$	$92 \div 4$	$110 \div 5$
# 5	$56 \div 2$	$87 \div 3$	$84 \div 4$	$100 \div 5$
# 6	$42 \div 2$	$60 \div 3$	$112 \div 4$	$145 \div 5$
# 7	$46 \div 2$	$66 \div 3$	$104 \div 4$	$135 \div 5$
# 8	$50 \div 2$	$72 \div 3$	$96 \div 4$	$125 \div 5$
# 9	$54 \div 2$	$78 \div 3$	$88 \div 4$	$115 \div 5$
# 10	$58 \div 2$	$84 \div 3$	$80 \div 4$	$105 \div 5$

Divide numbers in column

$9 \times 1 = 9$ $\quad 9 \times 2 = 18$ $\quad 9 \times 3 = 27$

$$\begin{array}{r} 29 \\ 9\overline{)261} \\ \underline{-18} \\ 81 \\ \underline{-81} \\ 0 \end{array}$$

1) $9 \times 2 = 18$
2) $26 - 18 = 8$
3) Bring down the 1
4) $9 \times 9 = 81$
5) $81 - 81 = 0$
6) $R = 0$

or

$$\begin{array}{rl} 261 & \underline{|\,9} \\ -\underline{18} & 29 \\ 81 & \\ \underline{-81} & \\ 0 & \end{array}$$

Divide numbers

	a	b	c	d
# 11	$120 \div 6$	$147 \div 7$	$232 \div 8$	$252 \div 9$
# 12	$132 \div 6$	$161 \div 7$	$216 \div 8$	$234 \div 9$
# 13	$144 \div 6$	$175 \div 7$	$200 \div 8$	$216 \div 9$
# 14	$156 \div 6$	$189 \div 7$	$184 \div 8$	$198 \div 9$
# 15	$168 \div 6$	$203 \div 7$	$168 \div 8$	$180 \div 9$
# 16	$126 \div 6$	$140 \div 7$	$224 \div 8$	$261 \div 9$
# 17	$138 \div 6$	$154 \div 7$	$208 \div 8$	$243 \div 9$
# 18	$150 \div 6$	$168 \div 7$	$192 \div 8$	$225 \div 9$
# 19	$162 \div 6$	$182 \div 7$	$176 \div 8$	$207 \div 9$
# 20	$174 \div 6$	$196 \div 7$	$160 \div 8$	$189 \div 9$

Summary 4 Chapter 3 Lesson 5
Division numbers with quotient from 30 to 39
Divide numbers in column 5
Divide numbers

	a	b	c	d
# 1	60 ÷ 2	93 ÷ 3	156 ÷ 4	190 ÷ 5
# 2	64 ÷ 2	99 ÷ 3	148 ÷ 4	180 ÷ 5
# 3	68 ÷ 2	105 ÷ 3	140 ÷ 4	170 ÷ 5
# 4	72 ÷ 2	111 ÷ 3	132 ÷ 4	160 ÷ 5
# 5	76 ÷ 2	117 ÷ 3	124 ÷ 4	150 ÷ 5
# 6	62 ÷ 2	90 ÷ 3	152 ÷ 4	195 ÷ 5
# 7	66 ÷ 2	96 ÷ 3	144 ÷ 4	185 ÷ 5
# 8	70 ÷ 2	102 ÷ 3	136 ÷ 4	175 ÷ 5
# 9	74 ÷ 2	108 ÷ 3	128 ÷ 4	165 ÷ 5
# 10	78 ÷ 2	114 ÷ 3	120 ÷ 4	155 ÷ 5

Divide numbers in column

39	1) 9× 3 = 27
9 ⟌351	2) 35 – 27 = 8
- 27	3) Bring down the 1
81	4) 9 × 9 = 81
- 81	5) 81 – 81 = 0
0	6) R = 0

Divide numbers

	a	b	c	d
# 11	180 ÷ 6	217 ÷ 7	312 ÷ 8	342 ÷ 9
# 12	192 ÷ 6	231 ÷ 7	296 ÷ 8	324 ÷ 9
# 13	204 ÷ 6	245 ÷ 7	280 ÷ 8	306 ÷ 9
# 14	216 ÷ 6	259 ÷ 7	264 ÷ 8	288 ÷ 9
# 15	228 ÷ 6	273 ÷ 7	248 ÷ 8	270 ÷ 9
# 16	186 ÷ 6	210 ÷ 7	304 ÷ 8	351 ÷ 9
# 17	198 ÷ 6	224 ÷ 7	288 ÷ 8	333 ÷ 9
# 18	210 ÷ 6	238 ÷ 7	272 ÷ 8	315 ÷ 9
# 19	222 ÷ 6	252 ÷ 7	256 ÷ 8	297 ÷ 9
# 20	234 ÷ 6	266 ÷ 7	240 ÷ 8	279 ÷ 9

Summary 4 Chapter 3 Lesson 6
Division numbers with quotient from 40 to 49
Divide numbers in column 6
Divide numbers

	a	b	c	d
# 1	80 ÷ 2	123 ÷ 3	196 ÷ 4	240 ÷ 5
# 2	84 ÷ 2	129 ÷ 3	188 ÷ 4	230 ÷ 5
# 3	88 ÷ 2	135 ÷ 3	180 ÷ 4	220 ÷ 5
# 4	92 ÷ 2	141 ÷ 3	172 ÷ 4	210 ÷ 5
# 5	96 ÷ 2	147 ÷ 3	164 ÷ 4	200 ÷ 5
# 6	82 ÷ 2	120 ÷ 3	192 ÷ 4	245 ÷ 5
# 7	86 ÷ 2	126 ÷ 3	184 ÷ 4	235 ÷ 5
# 8	90 ÷ 2	132 ÷ 3	176 ÷ 4	225 ÷ 5
# 9	94 ÷ 2	138 ÷ 3	168 ÷ 4	215 ÷ 5
# 10	98 ÷ 2	144 ÷ 3	160 ÷ 4	205 ÷ 5

Divide numbers in column

```
   49
9 |441
 - 36
    81
  - 81
```

1) 9 × 4 = 36
2) 44 – 36 = 8
3) Bring down the 1
4) 9× 9 = 81
5) 81 – 81 = 0

Divide numbers

	a	b	c	d
# 11	240 ÷ 6	287 ÷ 7	392 ÷ 8	432 ÷ 9
# 12	252 ÷ 6	301 ÷ 7	376 ÷ 8	414 ÷ 9
# 13	264 ÷ 6	315 ÷ 7	360 ÷ 8	396 ÷ 9
# 14	276 ÷ 6	329 ÷ 7	344 ÷ 8	378 ÷ 9
# 15	288 ÷ 6	343 ÷ 7	328 ÷ 8	360 ÷ 9
# 16	246 ÷ 6	280 ÷ 7	384 ÷ 8	441 ÷ 9
# 17	258 ÷ 6	294 ÷ 7	368 ÷ 8	423 ÷ 9
# 18	270 ÷ 6	308 ÷ 7	352 ÷ 8	405 ÷ 9
# 19	282 ÷ 6	322 ÷ 7	336 ÷ 8	387 ÷ 9
# 20	294 ÷ 6	336 ÷ 7	320 ÷ 8	369 ÷ 9

Summary 4 Chapter 3 Lesson 7
Division numbers with quotient from 50 to 59
Divide numbers in column 7

```
   59
5 |295
 - 25
    45
  - 45
```

1) 5×5 = 25
2) 29 – 25 = 4
3) Bring down the 5
4) 5×9 = 45
5) 45 – 45 = 0

Divide numbers

	a	b	c	d
# 1	100 ÷ 2	153 ÷ 3	236 ÷ 4	290 ÷ 5
# 2	104 ÷ 2	159 ÷ 3	228 ÷ 4	280 ÷ 5
# 3	108 ÷ 2	165 ÷ 3	220 ÷ 4	270÷ 5
# 4	112 ÷ 2	171 ÷ 3	212 ÷ 4	260 ÷ 5
# 5	116 ÷ 2	177 ÷ 3	204 ÷ 4	250 ÷ 5
# 6	102 ÷ 2	150 ÷ 3	232 ÷ 4	295 ÷ 5
# 7	106 ÷ 2	156 ÷ 3	224 ÷ 4	285 ÷ 5
# 8	110 ÷ 2	162 ÷ 3	216 ÷ 4	275 ÷ 5
# 9	114 ÷ 2	168 ÷ 3	208 ÷ 4	265 ÷ 5
# 10	118 ÷ 2	174 ÷ 3	200 ÷ 4	255 ÷ 5

Divide numbers in column

```
   59
9 |531
 - 45
    81
  - 81
```

1) 9 × 5 = 45
2) 53 – 45 = 8
3) Bring down the 1
4) 9 × 9 = 81
5) 81 – 81 = 0

Divide numbers

	a	b	c	d
# 11	300 ÷ 6	357 ÷ 7	472 ÷ 8	522 ÷ 9
# 12	312 ÷ 6	371 ÷ 7	456 ÷ 8	504 ÷ 9
# 13	324 ÷ 6	385 ÷ 7	440 ÷ 8	486 ÷ 9
# 14	336 ÷ 6	399 ÷ 7	424 ÷ 8	468 ÷ 9
# 15	348 ÷ 6	413 ÷ 7	408 ÷ 8	450 ÷ 9
# 16	306 ÷ 6	350 ÷ 7	464 ÷ 8	531 ÷ 9
# 17	318 ÷ 6	364 ÷ 7	448 ÷ 8	513 ÷ 9
# 18	330 ÷ 6	378 ÷ 7	432 ÷ 8	495 ÷ 9
# 19	342 ÷ 6	392 ÷ 7	416 ÷ 8	477 ÷ 9
# 20	354 ÷ 6	406 ÷ 7	400 ÷ 8	459 ÷ 9

Summary 4 Chapter 3 Lesson 8
Division numbers with quotient from 60 – 69
Divide numbers in column 8

$$\begin{array}{r} 69 \\ 5\,\overline{)345} \\ \underline{-30} \\ 45 \\ \underline{-45} \end{array}$$

1) $5 \times 6 = 30$
2) $34 - 30 = 4$
3) Bring down the 5
4) $5 \times 9 = 45$
5) $45 - 45 = 0$

Divide numbers

	a	b	c	d
# 1	120 ÷ 2	183 ÷ 3	276 ÷ 4	340 ÷ 5
# 2	124 ÷ 2	189 ÷ 3	268 ÷ 4	330 ÷ 5
# 3	128 ÷ 2	195 ÷ 3	260 ÷ 4	320 ÷ 5
# 4	132 ÷ 2	201 ÷ 3	252 ÷ 4	310 ÷ 5
# 5	136 ÷ 2	207 ÷ 3	244 ÷ 4	300 ÷ 5
# 6	122 ÷ 2	180 ÷ 3	272 ÷ 4	345 ÷ 5
# 7	126 ÷ 2	186 ÷ 3	264 ÷ 4	335 ÷ 5
# 8	130 ÷ 2	192 ÷ 3	256 ÷ 4	325 ÷ 5
# 9	134 ÷ 2	198 ÷ 3	248 ÷ 4	315 ÷ 5
# 10	138 ÷ 2	204 ÷ 3	240 ÷ 4	305 ÷ 5

Divide numbers in column

$$\begin{array}{r} 69 \\ 9\,\overline{)621} \\ \underline{-54} \\ 81 \\ \underline{-81} \end{array}$$

1) $9 \times 6 = 54$
2) $62 - 54 = 8$
3) Bring down the 1
4) $9 \times 9 = 81$
5) $81 - 81 = 0$

Divide numbers

	a	b	c	d
# 11	360 ÷ 6	427 ÷ 7	552 ÷ 8	612 ÷ 9
# 12	372 ÷ 6	441 ÷ 7	536 ÷ 8	594 ÷ 9
# 13	384 ÷ 6	455 ÷ 7	520 ÷ 8	576 ÷ 9
# 14	396 ÷ 6	469 ÷ 7	504 ÷ 8	558 ÷ 9
# 15	408 ÷ 6	483 ÷ 7	488 ÷ 8	540 ÷ 9
# 16	366 ÷ 6	420 ÷ 7	544 ÷ 8	621 ÷ 9
# 17	378 ÷ 6	434 ÷ 7	528 ÷ 8	603 ÷ 9
# 18	390 ÷ 6	448 ÷ 7	512 ÷ 8	585 ÷ 9
# 19	402 ÷ 6	462 ÷ 7	496 ÷ 8	567 ÷ 9
# 20	414 ÷ 6	476 ÷ 7	480 ÷ 8	549 ÷ 9

Summary 4 Chapter 3 Lesson 9
Division numbers with quotient from 70 – 79
Divide numbers in column 9
Divide numbers

	a	b	c	d
# 1	140 ÷ 2	213 ÷ 3	316 ÷ 4	390 ÷ 5
# 2	144 ÷ 2	219 ÷ 3	308 ÷ 4	380 ÷ 5
# 3	148 ÷ 2	225 ÷ 3	300 ÷ 4	370 ÷ 5
# 4	152 ÷ 2	231 ÷ 3	292 ÷ 4	360 ÷ 5
# 5	156 ÷ 2	237 ÷ 3	284 ÷ 4	350 ÷ 5
# 6	142 ÷ 2	210 ÷ 3	312 ÷ 4	395 ÷ 5
# 7	146 ÷ 2	216 ÷ 3	304 ÷ 4	385 ÷ 5
# 8	150 ÷ 2	222 ÷ 3	296 ÷ 4	375 ÷ 5
# 9	154 ÷ 2	228 ÷ 3	288 ÷ 4	365 ÷ 5
# 10	158 ÷ 2	234 ÷ 3	280 ÷ 4	355 ÷ 5

Divide numbers in column

```
   79
9 |711
 -63
   81
 - 81
```

1) 9× 7 = 63
2) 71 – 63 = 8
3) Bring down the 1
4) 9 × 9 = 81
5) 81 – 81 = 0

Divide numbers

	a	b	c	d
# 11	420 ÷ 6	497 ÷ 7	632 ÷ 8	702 ÷ 9
# 12	432 ÷ 6	511 ÷ 7	616 ÷ 8	684 ÷ 9
# 13	444 ÷ 6	525 ÷ 7	600 ÷ 8	666 ÷ 9
# 14	456 ÷ 6	539 ÷ 7	584 ÷ 8	648 ÷ 9
# 15	468 ÷ 6	553 ÷ 7	568 ÷ 8	630 ÷ 9
# 16	426 ÷ 6	490 ÷ 7	624 ÷ 8	711 ÷ 9
# 17	438 ÷ 6	504 ÷ 7	608 ÷ 8	693 ÷ 9
# 18	450 ÷ 6	518 ÷ 7	592 ÷ 8	675 ÷ 9
# 19	462 ÷ 6	532 ÷ 7	576 ÷ 8	657 ÷ 9
# 20	474 ÷ 6	546 ÷ 7	560 ÷ 8	639 ÷ 9

Summary 4 Chapter 3 Lesson 10
Division numbers with quotient from 80 to 89
Divide numbers in column 10
Divide numbers

	a	b	c	d
# 1	160 ÷ 2	243 ÷ 3	356 ÷ 4	440 ÷ 5
# 2	164 ÷ 2	249 ÷ 3	348 ÷ 4	430 ÷ 5
# 3	168 ÷ 2	255 ÷ 3	340 ÷ 4	420 ÷ 5
# 4	172 ÷ 2	261 ÷ 3	332 ÷ 4	410 ÷ 5
# 5	176 ÷ 2	267 ÷ 3	324 ÷ 4	400 ÷ 5
# 6	162 ÷ 2	240 ÷ 3	352 ÷ 4	445 ÷ 5
# 7	166 ÷ 2	246 ÷ 3	344 ÷ 4	435 ÷ 5
# 8	170 ÷ 2	252 ÷ 3	336 ÷ 4	425 ÷ 5
# 9	174 ÷ 2	258 ÷ 3	328 ÷ 4	415 ÷ 5
# 10	178 ÷ 2	264 ÷ 3	320 ÷ 4	405 ÷ 5

Divide numbers in column

```
   89
9 |801
 -72
   81
  -81
```

1) 9× 8 = 72
2) 80 – 72 = 8
3) Bring down the 1
4) 9 × 9 = 81
5) 81 – 81 = 0

Divide numbers

	a	b	c	d
# 11	480 ÷ 6	567 ÷ 7	712 ÷ 8	792 ÷ 9
# 12	492 ÷ 6	581 ÷ 7	696 ÷ 8	774 ÷ 9
# 13	504 ÷ 6	595 ÷ 7	680 ÷ 8	756 ÷ 9
# 14	516 ÷ 6	609 ÷ 7	664 ÷ 8	738 ÷ 9
# 15	528 ÷ 6	623 ÷ 7	648 ÷ 8	720 ÷ 9
# 16	486 ÷ 6	560 ÷ 7	704 ÷ 8	801 ÷ 9
# 17	498 ÷ 6	574 ÷ 7	688 ÷ 8	783 ÷ 9
# 18	510 ÷ 6	588 ÷ 7	672 ÷ 8	765 ÷ 9
# 19	522 ÷ 6	602 ÷ 7	656 ÷ 8	747 ÷ 9
# 20	534 ÷ 6	616 ÷ 7	640 ÷ 8	729 ÷ 9

Summary 4 Chapter 3 Lesson 11
Division numbers with quotient from 90 to 99
Divide numbers in column 11
Divide numbers

	a	b	c	d
# 1	180 ÷ 2	273 ÷ 3	396 ÷ 4	490 ÷ 5
# 2	184 ÷ 2	279 ÷ 3	388 ÷ 4	480 ÷ 5
# 3	188 ÷ 2	285 ÷ 3	380 ÷ 4	470 ÷ 5
# 4	192 ÷ 2	291 ÷ 3	372 ÷ 4	460 ÷ 5
# 5	196 ÷ 2	297 ÷ 3	364 ÷ 4	450 ÷ 5
# 6	182 ÷ 2	270 ÷ 3	392 ÷ 4	495 ÷ 5
# 7	186 ÷ 2	276 ÷ 3	384 ÷ 4	485 ÷ 5
# 8	190 ÷ 2	282 ÷ 3	376 ÷ 4	475 ÷ 5
# 9	194 ÷ 2	288 ÷ 3	368 ÷ 4	465 ÷ 5
# 10	198 ÷ 2	294 ÷ 3	360 ÷ 4	455 ÷ 5

Divide numbers in column

```
   97
8 |776
  -72
   56
  -56
```

1) $8 \times 9 = 72$
2) $77 - 72 = 5$
3) Bring down the 6
4) $8 \times 7 = 56$
5) $56 - 56 = 0$

Divide numbers

	a	b	c	d
# 11	540 ÷ 6	637 ÷ 7	792 ÷ 8	882 ÷ 9
# 12	552 ÷ 6	651 ÷ 7	776 ÷ 8	864 ÷ 9
# 13	564 ÷ 6	665 ÷ 7	760 ÷ 8	846 ÷ 9
# 14	576 ÷ 6	679 ÷ 7	744 ÷ 8	828 ÷ 9
# 15	588 ÷ 6	693 ÷ 7	728 ÷ 8	810 ÷ 9
# 16	546 ÷ 6	630 ÷ 7	784 ÷ 8	891 ÷ 9
# 17	558 ÷ 6	644 ÷ 7	768 ÷ 8	873 ÷ 9
# 18	570 ÷ 6	658 ÷ 7	752 ÷ 8	855 ÷ 9
# 19	582 ÷ 6	672 ÷ 7	736 ÷ 8	837 ÷ 9
# 20	594 ÷ 6	686 ÷ 7	720 ÷ 8	819 ÷ 9

Summary 4 Chapter 3 Lesson 12
Division numbers
Review 1 A
Divide the numbers in column

# 1	490 ÷5	# 11	531 ÷ 9
# 2	522 ÷ 6	# 12	544 ÷ 8
# 3	532 ÷ 7	# 13	525 ÷ 7
# 4	520 ÷ 8	# 14	516 ÷ 6
# 5	486 ÷ 9	# 15	475 ÷ 5
# 6	485 ÷ 5	# 16	504 ÷ 9
# 7	516 ÷ 6	# 17	536 ÷ 8
# 8	525 ÷ 7	# 18	546 ÷ 7
# 9	512 ÷ 8	# 19	534 ÷ 6
# 10	477 ÷ 9	# 20	480 ÷ 5

Review 1 B
Divide the numbers in column

# 1	395 ÷5	# 11	472 ÷ 8
# 2	602 ÷ 7	# 12	552 ÷ 8
# 3	608 ÷ 8	# 13	450 ÷ 6
# 4	536 ÷ 8	# 14	696 ÷ 8
# 5	504 ÷ 9	# 15	582 ÷ 6
# 6	390 ÷ 5	# 16	448 ÷ 8
# 7	595 ÷ 7	# 17	603 ÷ 9
# 8	450 ÷ 6	# 18	702 ÷ 9
# 9	576 ÷ 9	# 19	623 ÷ 7
# 10	424 ÷ 8	# 20	445 ÷ 5

Summary 4 Chapter 3 Lesson 13

Review 2 A

Divide the numbers in column

# 1	480 ÷ 5	# 11	513 ÷ 9
# 2	510 ÷ 6	# 12	528 ÷ 8
# 3	518 ÷ 7	# 13	504 ÷ 7
# 4	504 ÷ 8	# 14	504 ÷ 6
# 5	468 ÷ 9	# 15	465 ÷ 5
# 6	475 ÷ 5	# 16	486 ÷ 9
# 7	504 ÷ 6	# 17	520 ÷ 8
# 8	511 ÷ 7	# 18	532 ÷ 7
# 9	496 ÷ 8	# 19	522 ÷ 6
# 10	459 ÷ 9	# 20	470 ÷ 5

Review 2 B

Divide the numbers in column

# 1	385 ÷5	# 11	456 ÷ 8
# 2	588 ÷ 7	# 12	536 ÷ 8
# 3	592 ÷ 8	# 13	438 ÷ 6
# 4	520 ÷ 8	# 14	680 ÷ 8
# 5	486 ÷ 9	# 15	570 ÷ 6
# 6	380 ÷ 5	# 16	432 ÷ 8
# 7	581 ÷ 7	# 17	585 ÷ 9
# 8	438 ÷ 6	# 18	684 ÷ 9
# 9	558 ÷ 9	# 19	609 ÷ 7
# 10	408 ÷ 8	# 20	430 ÷ 5

Summary 4 Chapter 3 Lesson 14
Summary 3

1) Multiplication numbers: for memory (x, x, any)
To multiply numbers, work in any order
$9 \times 8 \times 7 = 72 \times 7 = 504$ $9 \times 8 \times 7 = 9 \times 56 = 504$

2) Work from left to right: for memory (L → R)

a) Multiplication and Division numbers: (× or ÷)
To multiply and divide numbers, work from left to right (L → R)
$9 \times 6 \div 3 = 54 \div 3 = 18$ $4 \times 81 \div 9 = 324 \div 9 = 36$

b) Division and Multiplication numbers: (÷ or ×)
To divide and multiply numbers, work from left to right (L → R)
$24 \div 6 \times 2 = 4 \times 2 = 8$ $81 \div 9 \times 4 = 9 \times 4 = 36$

c) Division numbers: (÷, ÷)
To divide numbers, work from left to right (L → R)
$24 \div 4 \div 2 = 6 \div 2 = 3$ $810 \div 9 \div 3 = 90 \div 9 = 10$

3) Do Multiplication or Division before Addition or Subtraction for memory: (× or ÷) before (+ or -)
a) Multiplication or Addition numbers: 1) × 2) +
To multiply or add numbers, always first multiply than add
$9 \times 6 + 9 = 54 + 9 = 63$ $9 + 9 \times 9 = 9 + 81 = 90$

b) Multiplication or Subtraction numbers: 1) × 2) −
To multiply or subtract numbers, always first multiply than subtract
$9 \times 9 - 9 = 81 - 9 = 72$ $90 - 9 \times 9 = 90 - 81 = 9$

c) Division or Addition numbers: 1) ÷ 2) +
To divide or add numbers, always first divide than add
$50 \div 5 + 39 = 10 + 39 = 49$ $88 + 171 \div 9 = 88 + 19 = 107$

d) Division or Subtraction numbers: 1) ÷ 2) −
To divide or subtract numbers, always first divide than subtract
$108 \div 6 - 9 = 18 - 9 = 9$ $500 - 126 \div 7 = 500 - 18 = 482$

Summary 4 Chapter 3 Lesson 14

Multiplication numbers

To multiply numbers, work in any order

$5\times6\times10=30\times10=300$

$5\times6\times10 = 5\times60 = 300$

Multiply numbers. Review 1 A

	a	b	c
# 1	2 × 2 × 1	3 × 4 × 1	4 × 6 × 1
# 2	2 × 2 × 2	3 × 4 × 2	4 × 6 × 2
# 3	2 × 2 × 3	3 × 4 × 3	4 × 6 × 3
# 4	2 × 2 × 4	3 × 4 × 4	4 × 6 × 4
# 5	2 × 2 × 5	3 × 4 × 5	4 × 6 × 5
# 6	2 × 2 × 6	3 × 4 × 6	4 × 6 × 6
# 7	2 × 2 × 7	3 × 4 × 7	4 × 6 × 7
# 8	2 × 2 × 8	3 × 4 × 8	4 × 6 × 8
# 9	2 × 2 × 9	3 × 4 × 9	4 × 6 × 9
# 10	2 × 2 × 10	3 × 4 × 10	4 × 6 × 10

To multiply numbers, work in any order

$5\times8\times10 =40\times10=400$

$5\times8\times10 = 5\times80 = 400$

Multiply numbers

	a	b	c
# 11	5 × 3 × 1	6 × 5 × 1	7 × 7 × 1
# 12	5 × 3 × 2	6 × 5 × 2	7 × 7 × 2
# 13	5 × 3 × 3	6 × 5 × 3	7 × 7× 3
# 14	5 × 3 × 4	6 × 5 × 4	7 × 7 × 4
# 15	5 × 3 × 5	6 × 5 × 5	7 × 7 × 5
# 16	5 × 3 × 6	6 × 5 × 6	7 × 7 × 6
# 17	5 × 3 × 7	6 × 5 × 7	7 × 7 × 7
# 18	5 × 3 × 8	6 × 5 × 8	7 × 7 × 8
# 19	5 × 3 × 9	6 × 5 × 9	7 × 7 × 9
# 20	5 × 3 × 10	6 × 5 × 10	7 × 7 × 10

Summary 4 Chapter 3 Lesson 14

Multiplication and Division numbers

To multiply and divide numbers work from left to right

$$5\times 60\div 6 = 300\div 6 = 50$$

Multiply numbers. Review 1 B

	a	b	c
# 1	$2\times 2\div 2$	$3\times 4\div 4$	$4\times 6\div 6$
# 2	$2\times 4\div 2$	$3\times 8\div 4$	$4\times 12\div 6$
# 3	$2\times 6\div 2$	$3\times 12\div 4$	$4\times 18\div 6$
# 4	$2\times 8\div 2$	$3\times 16\div 4$	$4\times 24\div 6$
# 5	$2\times 10\div 2$	$3\times 20\div 4$	$4\times 30\div 6$
# 6	$2\times 12\div 2$	$3\times 24\div 4$	$4\times 36\div 6$
# 7	$2\times 14\div 2$	$3\times 28\div 4$	$4\times 42\div 6$
# 8	$2\times 16\div 2$	$3\times 32\div 4$	$4\times 48\div 6$
# 9	$2\times 18\div 2$	$3\times 36\div 4$	$4\times 54\div 6$
# 10	$2\times 20\div 2$	$3\times 40\div 4$	$4\times 60\div 6$

Multiplication and Division numbers

To multiply and divide numbers work from left to right

$$5\times 80\div 8 = 400\div 8 = 50$$

Multiply numbers

	a	b	c
# 11	$5\times 3\div 3$	$6\times 5\div 5$	$7\times 7\div 7$
# 12	$5\times 6\div 3$	$6\times 10\div 5$	$7\times 14\div 7$
# 13	$5\times 9\div 3$	$6\times 15\div 5$	$7\times 21\div 7$
# 14	$5\times 12\div 3$	$6\times 20\div 5$	$7\times 28\div 7$
# 15	$5\times 15\div 3$	$6\times 25\div 5$	$7\times 35\div 7$
# 16	$5\times 18\div 3$	$6\times 30\div 5$	$7\times 42\div 7$
# 17	$5\times 21\div 3$	$6\times 35\div 5$	$7\times 49\div 7$
# 18	$5\times 24\div 3$	$6\times 40\div 5$	$7\times 56\div 7$
# 19	$5\times 27\div 3$	$6\times 45\div 5$	$7\times 63\div 7$
# 20	$5\times 30\div 3$	$6\times 50\div 5$	$7\times 70\div 7$

Summary 4 Chapter 3 Lesson 15

Division and Multiplication numbers

To divide and multiply numbers work from left to right

$$70 \div 7 \times 4 = 10 \times 4 = 40$$

Evaluate. Review 1 A

	a	b	c
# 1	$2 \div 2 \times 2$	$4 \div 4 \times 3$	$6 \div 6 \times 4$
# 2	$4 \div 2 \times 2$	$8 \div 4 \times 3$	$12 \div 6 \times 4$
# 3	$6 \div 2 \times 2$	$12 \div 4 \times 3$	$18 \div 6 \times 4$
# 4	$8 \div 2 \times 2$	$16 \div 4 \times 3$	$24 \div 6 \times 4$
# 5	$10 \div 2 \times 2$	$20 \div 4 \times 3$	$30 \div 6 \times 4$
# 6	$12 \div 2 \times 2$	$24 \div 4 \times 3$	$36 \div 6 \times 4$
# 7	$14 \div 2 \times 2$	$28 \div 4 \times 3$	$42 \div 6 \times 4$
# 8	$16 \div 2 \times 2$	$32 \div 4 \times 3$	$48 \div 6 \times 4$
# 9	$18 \div 2 \times 2$	$36 \div 4 \times 3$	$54 \div 6 \times 4$
# 10	$20 \div 2 \times 2$	$40 \div 4 \times 3$	$60 \div 6 \times 4$

Division and Multiplication numbers

To divide and multiply numbers work from left to right

$$80 \div 8 \times 8 = 10 \times 8 = 80$$

Evaluate

	a	b	c
# 11	$3 \div 3 \times 5$	$5 \div 5 \times 6$	$7 \div 7 \times 7$
# 12	$6 \div 3 \times 5$	$10 \div 5 \times 6$	$14 \div 7 \times 7$
# 13	$9 \div 3 \times 5$	$15 \div 5 \times 6$	$21 \div 7 \times 7$
# 14	$12 \div 3 \times 5$	$20 \div 5 \times 6$	$28 \div 7 \times 7$
# 15	$15 \div 3 \times 5$	$25 \div 5 \times 6$	$35 \div 7 \times 7$
# 16	$18 \div 3 \times 5$	$30 \div 5 \times 6$	$42 \div 7 \times 7$
# 17	$21 \div 3 \times 5$	$35 \div 5 \times 6$	$49 \div 7 \times 7$
# 18	$24 \div 3 \times 5$	$40 \div 5 \times 6$	$56 \div 7 \times 7$
# 19	$27 \div 3 \times 5$	$45 \div 5 \times 6$	$63 \div 7 \times 7$
# 20	$30 \div 3 \times 5$	$50 \div 5 \times 6$	$70 \div 7 \times 7$

Summary 4 Chapter 3 Lesson 15

Division numbers

To divide numbers, work from left to right

640 ÷ 8 ÷ 8 = 80 ÷ 8 = 10

Divide numbers. Review 1 B

	a	b	c
# 1	4 ÷ 2 ÷ 2	16 ÷ 4 ÷ 4	36 ÷ 6 ÷ 6
# 2	8 ÷ 2 ÷ 2	32 ÷ 4 ÷ 4	72 ÷ 6÷ 6
# 3	12 ÷ 2 ÷ 2	48 ÷ 4 ÷ 4	108 ÷ 6 ÷ 6
# 4	16 ÷ 2 ÷ 2	64 ÷ 4 ÷ 4	144 ÷ 6 ÷ 6
# 5	20 ÷ 2 ÷ 2	80 ÷ 4 ÷ 4	180 ÷ 6 ÷ 6
# 6	24 ÷ 2 ÷ 2	96 ÷ 4 ÷ 4	216 ÷ 6 ÷ 6
# 7	28 ÷ 2 ÷ 2	112 ÷ 4 ÷ 4	252 ÷ 6 ÷ 6
# 8	32 ÷ 2 ÷ 2	128 ÷ 4÷ 4	288 ÷ 6 ÷ 6
# 9	36 ÷ 2 ÷ 2	144 ÷ 4 ÷ 4	324 ÷ 6 ÷ 6
# 10	40 ÷ 2 ÷ 2	160 ÷ 4 ÷ 4	360 ÷ 6 ÷ 6

Divide numbers

To divide numbers, work from left to right

810 ÷ 9 ÷ 9 = 90 ÷ 9 = 10

Divide numbers

	a	b	c
# 11	9 ÷ 3 ÷ 3	25÷5 ÷ 5	49 ÷ 7÷7
# 12	18 ÷ 3 ÷3	50 ÷5÷5	98 ÷7÷7
# 13	27÷3 ÷3	75 ÷5÷5	147 ÷7÷7
# 14	36 ÷3÷3	100 ÷5÷5	196 ÷7÷7
# 15	45 ÷3÷3	125 ÷5÷5	245÷7 ÷7
# 16	54 ÷3÷3	150 ÷5÷5	294 ÷7 ÷7
# 17	63 ÷3÷3	175÷5÷5	343÷7÷ 7
# 18	72 ÷3÷3	200÷5÷5	392 ÷7÷ 7
# 19	81 ÷3÷3	225÷5 ÷5	441 ÷7÷ 7
# 20	90 ÷3÷3	250÷5÷5	490 ÷7÷ 7

Summary 4 Chapter 3 Lesson 16

Division and Addition numbers

To divide and add numbers always first divide than add

$$50 \div 5 + 10 = 10 + 10 = 20$$

Evaluate. Review 1 A

	a	b	c	d
# 1	$2 \div 2 + 0$	$3 \div 3 + 0$	$4 \div 4 + 0$	$5 \div 5 + 0$
# 2	$4 \div 2 + 1$	$6 \div 3 + 1$	$8 \div 4 + 1$	$10 \div 5 + 1$
# 3	$6 \div 2 + 2$	$9 \div 3 + 2$	$12 \div 4 + 2$	$15 \div 5 + 2$
# 4	$8 \div 2 + 3$	$12 \div 3 + 3$	$16 \div 4 + 3$	$20 \div 5 + 3$
# 5	$10 \div 2 + 4$	$15 \div 3 + 4$	$20 \div 4 + 4$	$25 \div 5 + 4$
# 6	$12 \div 2 + 5$	$18 \div 3 + 5$	$24 \div 4 + 5$	$30 \div 5 + 5$
# 7	$14 \div 2 + 6$	$21 \div 3 + 6$	$28 \div 4 + 6$	$35 \div 5 + 6$
# 8	$16 \div 2 + 7$	$24 \div 3 + 7$	$32 \div 4 + 7$	$40 \div 5 + 7$
# 9	$18 \div 2 + 8$	$27 \div 3 + 8$	$36 \div 4 + 8$	$45 \div 5 + 8$
# 10	$20 \div 2 + 9$	$30 \div 3 + 9$	$40 \div 4 + 9$	$50 \div 5 + 9$

Divide and Add

$$10 \overset{2}{+} 90 \overset{1}{\div} 9 = 10 + 10 = 20$$

Evaluate

	a	b	c	d
# 11	$0 + 6 \div 6$	$0 + 7 \div 7$	$0 + 8 \div 8$	$0 + 9 \div 9$
# 12	$1 + 12 \div 6$	$1 + 14 \div 7$	$1 + 16 \div 8$	$1 + 18 \div 9$
# 13	$2 + 18 \div 6$	$2 + 21 \div 7$	$2 + 24 \div 8$	$2 + 27 \div 9$
# 14	$3 + 24 \div 6$	$3 + 28 \div 7$	$3 + 32 \div 8$	$3 + 36 \div 9$
# 15	$4 + 30 \div 6$	$4 + 35 \div 7$	$4 + 40 \div 8$	$4 + 45 \div 9$
# 16	$5 + 36 \div 6$	$5 + 42 \div 7$	$5 + 48 \div 8$	$5 + 54 \div 9$
# 17	$6 + 42 \div 6$	$6 + 49 \div 7$	$6 + 56 \div 8$	$6 + 63 \div 9$
# 18	$7 + 48 \div 6$	$7 + 56 \div 7$	$7 + 64 \div 8$	$7 + 72 \div 9$
# 19	$8 + 54 \div 6$	$8 + 63 \div 7$	$8 + 72 \div 8$	$8 + 81 \div 9$
# 20	$9 + 60 \div 6$	$9 + 70 \div 7$	$9 + 80 \div 8$	$9 + 90 \div 9$

Summary 4 Chapter 3 Lesson 16

Division and Subtraction numbers

To divide and subtract numbers always first divide than subtract

80÷5 - 10 = 16 - 10 = 6

Evaluate. Review 1 B

	a	b	c	d
# 1	22÷2 - 1	24÷3 - 2	24÷4 - 3	25÷5 - 4
# 2	24÷2 - 1	27÷3- 2	28÷4 - 3	30÷5 - 4
# 3	26÷2 - 1	30÷3 –2	32÷4 - 3	35÷5- 4
# 4	28÷2 - 1	33÷3 - 2	36÷4 - 3	40÷5 - 4
# 5	30÷2 - 1	36÷3 - 2	40÷4 - 3	45÷5 - 4
# 6	32÷2 - 1	39÷3 - 2	44÷4 - 3	50÷5 - 4
# 7	34÷2 - 1	42÷3 - 2	48÷4 - 3	55÷5 - 4
# 8	36÷2 - 1	45÷3 - 2	52÷4 – 3	60÷5 - 4
# 9	38÷2 - 1	48÷3 - 2	56÷4 - 3	65÷5 - 4
# 10	40÷2 – 1	51÷3 – 2	60÷4 – 3	70÷5 –4

Divide and Subtract numbers

2 1

50 - 20 ÷4 = 50 - 5 = 45

Evaluate

	a	b	c	d
# 11	18 - 6÷6	21 - 7÷7	24 - 8÷8	27 - 9÷9
# 12	24 - 12÷6	28 - 14÷7	32 - 16÷8	36 - 18÷9
# 13	30 - 18÷6	35 - 21÷7	36 - 24÷8	45 - 27÷9
# 14	36 - 24÷6	42 - 28÷7	48 - 32÷8	54 - 36÷9
# 15	42 - 30÷6	49 - 35÷7	56 - 40÷8	63 - 45÷9
# 16	48 - 36÷6	56 - 42÷7	64 - 48÷8	72 - 54÷9
# 17	54 - 42÷6	63 - 49÷7	72 - 56÷8	81 - 63÷9
# 18	60 - 48÷6	70 - 56÷7	80 - 64÷8	90 - 72÷9
# 19	66 - 54÷6	77 - 63÷7	88 - 72÷8	99 - 81÷9
# 20	72 – 60÷6	84 –70÷7	96 –80÷8	108 – 90÷9

Summary 4 Chapter 3 Lesson 17

Review 2 A
Evaluate

1 $9 \times 2 \times 7$
2 $9 \times 18 \div 6$

3 $54 \div 9 \times 4$
4 $819 \div 7 \div 9$

5 $100 + 7 \times 7$
6 $80 \times 9 - 8$

7 $140 - 8 \times 5$
8 $800 \div 8 - 6$

9 $666 + 333 \div 3$
10 $840 - 240 \div 8$

11 $9 \times 3 \times 6$
12 $8 \times 27 \div 3$

13 $63 \div 7 \times 5$
14 $828 \div 4 \div 9$

15 $110 + 6 \times 6$
16 $90 \times 8 - 9$

17 $150 - 8 \times 6$
18 $840 \div 8 - 6$

19 $440 + 200 \div 5$
20 $870 - 350 \div 5$

Review 2 B
Evaluate

1 $9 \times 4 \times 5$
2 $6 \times 36 \div 9$

3 $63 \div 9 \times 3$
4 $108 \div 9 \div 3$

5 $104 + 8 \times 7$
6 $90 \times 8 - 20$

7 $120 - 12 \times 5$
8 $960 \div 8 - 20$

9 $260 + 280 \div 7$
10 $350 - 400 \div 8$

11 $9 \times 5 \times 4$
12 $7 \times 45 \div 9$

13 $54 \div 9 \times 2$
14 $972 \div 9 \div 2$

15 $101 + 7 \times 7$
16 $80 \times 7 - 60$

17 $170 - 14 \times 5$
18 $900 \div 9 - 60$

19 $270 + 180 \div 6$
20 $450 - 450 \div 9$

Summary 4 Chapter 3 Lesson 18
Four operations with numbers

Test 3 A

Evaluate

# 1	$9 \times 2 \times 9$	# 11	$9 \times 3 \times 8$
# 2	$9 \times 18 \div 9$	# 12	$8 \times 27 \div 9$
# 3	$54 \div 9 \times 5$	# 13	$63 \div 9 \times 4$
# 4	$819 \div 9 \div 7$	# 14	$828 \div 9 \div 4$
# 5	$100 + 9 \times 9$	# 15	$110 + 8 \times 8$
# 6	$80 \times 9 - 6$	# 16	$90 \times 8 - 7$
# 7	$140 - 9 \times 5$	# 17	$150 - 9 \times 6$
# 8	$840 \div 8 - 6$	# 18	$800 \div 8 - 6$
# 9	$666 + 333 \div 9$	# 19	$440 + 200 \div 8$
# 10	$840 - 240 \div 6$	# 20	$850 - 350 \div 7$

Summary 4 Chapter 3 Lesson 18

Four operations with numbers

Test 3 B

Evaluate

# 1	$9 \times 4 \times 7$	# 11	$9 \times 5 \times 6$
# 2	$7 \times 36 \div 9$	# 12	$6 \times 45 \div 9$
# 3	$72 \div 9 \times 3$	# 13	$81 \div 9 \times 2$
# 4	$837 \div 9 \div 3$	# 14	$846 \div 9 \div 2$
# 5	$120 + 8 \times 7$	# 15	$130 + 7 \times 7$
# 6	$90 \times 8 - 5$	# 16	$80 \times 9 - 6$
# 7	$160 - 9 \times 7$	# 17	$170 - 9 \times 8$
# 8	$960 \div 8 - 6$	# 18	$900 \div 9 - 6$
# 9	$210 + 280 \div 7$	# 19	$300 + 180 \div 6$
# 10	$800 - 400 \div 8$	# 20	$900 - 450 \div 9$

Summary 4 Chapter 3 Lesson 19
Review 4 A
Evaluate

# 1	$8 \times 5 \times 6$
# 2	$9 \times 24 \div 8$
# 3	$56 \div 8 \times 4$
# 4	$640 \div 8 \div 4$
# 5	$109 + 9 \times 9$
# 6	$60 \times 8 - 80$
# 7	$240 - 8 \times 5$
# 8	$720 \div 9 - 20$
# 9	$333 + 444 \div 4$
# 10	$360 - 360 \div 6$

Review 4 B
Evaluate

# 1	$6 \times 10 \times 10$
# 2	$7 \times 24 \div 6$
# 3	$72 \div 9 \times 4$
# 4	$480 \div 8 \div 4$
# 5	$104 + 6 \times 6$
# 6	$80 \times 9 - 20$
# 7	$135 - 7 \times 5$
# 8	$630 \div 9 - 10$
# 9	$444 + 555 \div 5$
# 10	$560 - 420 \div 7$

Summary 4 Chapter 3 Lesson 20

Final Exam 4 A

Evaluate

\# 1 $6 \times 5 \times 8$

\# 2 $8 \times 27 \div 9$

\# 3 $63 \div 9 \times 4$

\# 4 $720 \div 9 \div 4$

\# 5 $106 + 8 \times 8$

\# 6 $70 \times 9 - 30$

\# 7 $145 - 9 \times 5$

\# 8 $480 \div 8 - 20$

\# 9 $222 + 333 \div 3$

\# 10 $240 - 240 \div 6$

Summary 4 Chapter 3 Lesson 20

Final Exam 4 B
1 - # 10 all
Evaluate

\# 1 $8 \times 10 \times 6$

\# 2 $6 \times 45 \div 9$

\# 3 $81 \div 9 \times 2$

\# 4 $810 \div 9 \div 2$

\# 5 $111 + 7 \times 7$

\# 6 $90 \times 8 - 20$

\# 7 $160 - 8 \times 7$

\# 8 $640 \div 8 - 20$

\# 9 $210 + 280 \div 7$

\# 10 $250 - 400 \div 8$

Summary 4 Answers Chapter 3

Lesson 1

	a	b	c	d
# 1	1	6	1	6
# 2	2	7	2	7
# 3	3	8	3	8
# 4	4	9	4	9
# 5	5	10	5	10

	a	b	c	d
# 6	1	6	1	6
# 7	2	7	2	7
# 8	3	8	3	8
# 9	4	9	4	9
# 10	5	10	5	10

Lesson 2

	a	b	c	d	e	f	g	h	i
	×	2	3	4	5	6	7	8	9
# 1	1	2	3	4	5	6	7	8	9
# 2	2	4	6	8	10	12	14	16	18
# 3	3	6	9	12	15	18	21	24	27
# 4	4	8	12	16	20	24	28	32	36
# 5	5	10	15	20	25	30	35	40	45
# 6	6	12	18	24	30	36	42	48	54
# 7	7	14	21	28	35	42	49	56	63
# 8	8	16	24	32	40	48	56	64	72
# 9	9	18	27	36	45	54	63	72	81
#10	10	20	30	40	50	60	70	80	90

Lesson 3

	a b c d
# 1	10, 11, 19, 18
# 2	12, 13, 17, 16
# 3	14, 15, 15, 14
# 4	16, 17, 13, 12
# 5	18, 19, 11, 10
# 6	11, 10, 18, 19
# 7	13, 12, 16, 17
# 8	15, 14, 14, 15
# 9	17, 16, 12, 13
# 10	19, 18, 10, 11

	a b c d
# 11	10, 11, 19, 18
# 12	12, 13, 17, 16
# 13	14, 15, 15, 14
# 14	16, 17, 13, 12
# 15	18, 19, 11, 10
# 16	11, 10, 18, 19
# 17	13, 12, 16, 17
# 18	15, 14, 14, 15
# 19	17, 16, 12, 13
# 20	19, 18, 10, 11

Summary 4 Answers Chapter 3

Lesson 4

	a, b, c, d
# 1	20, 21, 29, 28
# 2	22, 23, 27, 26
# 3	24, 25, 25, 24
# 4	26, 27, 23, 22
# 5	28, 29, 21, 20
# 6	21, 20, 28, 29
# 7	23, 22, 26, 27
# 8	25, 24, 24, 25
# 9	27, 26, 22, 23
# 10	29, 28, 20, 21
# 11	20, 21, 29, 28
# 12	22, 23, 27, 26
# 13	24, 25, 25, 24
# 14	26, 27, 23, 22
# 15	28, 29, 21, 20
# 16	21, 20, 28, 29
# 17	23, 22, 26, 27
# 18	25, 24, 24, 25
# 19	27, 26, 22, 23
# 20	29, 28, 20, 21

Lesson 5

	a, b, c, d
# 1	30, 31, 39, 38
# 2	32, 33, 37, 36
# 3	34, 35, 35, 34
# 4	36, 37, 33, 32
# 5	38, 39, 31, 30
# 6	31, 30, 38, 39
# 7	33, 32, 36, 37
# 8	35, 34, 34, 35
# 9	37, 36, 32, 33
# 10	39, 38, 30, 31
# 11	30, 31, 39, 38
# 12	32, 33, 37, 36
# 13	34, 35, 35, 34
# 14	36, 37, 33, 32
# 15	38, 39, 31, 30
# 16	31, 30, 38, 39
# 17	33, 32, 36, 37
# 18	35, 34, 34, 35
# 19	37, 36, 32, 33
# 20	39, 38, 30, 31

Lesson 6

	a, b, c, d
# 1	40, 41, 49, 48
# 2	42, 43, 47, 46
# 3	44, 45, 45, 44
# 4	46, 47, 43, 42
# 5	48, 49, 41, 40
# 6	41, 40, 48, 49
# 7	43, 42, 46, 47
# 8	45, 44, 44, 45
# 9	47, 46, 42, 43
# 10	49, 48, 40, 41
# 11	40, 41, 49, 48
# 12	42, 43, 47, 46
# 13	44, 45, 45, 44
# 14	46, 47, 43, 42
# 15	48, 49, 41, 40
# 16	41, 40, 48, 49
# 17	43, 42, 46, 47
# 18	45, 44, 44, 45
# 19	47, 46, 42, 43
# 20	49, 48, 40, 41

Summary 4 Answers Chapter 3

Lesson 7

	a b c d		a b c d
# 1	50, 51, 59, 58	# 11	50, 51, 59, 58
# 2	52, 53, 57, 56	# 12	52, 53, 57, 56
# 3	54, 55, 55, 54	# 13	54, 55, 55, 54
# 4	56, 57, 53, 52	# 14	56, 57, 53, 52
# 5	58, 59, 51, 50	# 15	58, 59, 51, 50
# 6	51, 50, 58, 59	# 16	51, 50, 58, 59
# 7	53, 52, 56, 57	# 17	53, 52, 56, 57
# 8	55, 54, 54, 55	# 18	55, 54, 54, 55
# 9	57, 56, 52, 53	# 19	57, 56, 52, 53
# 10	59, 58, 50, 51	# 20	59, 58, 50, 51

Lesson 8

	a b c d		a b c d
# 1	60, 61, 69, 68	# 11	60, 61, 69, 68
# 2	62, 63, 67, 66	# 12	62, 63, 67, 66
# 3	64, 65, 65, 64	# 13	64, 65, 65, 64
# 4	66, 67, 63, 62	# 14	66, 67, 63, 62
# 5	68, 69, 61, 60	# 15	68, 69, 61, 60
# 6	61, 60, 68, 69	# 16	61, 60, 68, 69
# 7	63, 62, 66, 67	# 17	63, 62, 66, 67
# 8	65, 64, 64, 65	# 18	65, 64, 64, 65
# 9	67, 66, 62, 63	# 19	67, 66, 62, 63
# 10	69, 68, 60, 61	# 20	69, 68, 60, 61

Lesson 9

	a b c d		a b c d
# 1	70, 71, 79, 78	# 11	70, 71, 79, 78
# 2	72, 73, 77, 76	# 12	72, 73, 77, 76
# 3	74, 75, 75, 74	# 13	74, 75, 75, 74
# 4	76, 77, 73, 72	# 14	76, 77, 73, 72
# 5	78, 79, 71, 70	# 15	78, 79, 71, 70
# 6	71, 70, 78, 79	# 16	71, 70, 78, 79
# 7	73, 72, 76, 77	# 17	73, 72, 76, 77
# 8	75, 74, 74, 75	# 18	75, 74, 74, 75
# 9	77, 76, 72, 73	# 19	77, 76, 72, 73
# 10	79, 78, 70, 71	# 20	79, 78, 70, 71

Summary 4 Answers Chapter 3

Lesson 10

	a b c d		a b c d
# 1	80, 81, 89, 88	# 11	80, 81, 89, 88
# 2	82, 83, 87, 86	# 12	82, 83, 87, 86
# 3	84, 85, 85, 84	# 13	84, 85, 85, 84
# 4	86, 87, 83, 82	# 14	86, 87, 83, 82
# 5	88, 89, 81, 80	# 15	88, 89, 81, 80
# 6	81, 80, 88, 89	# 16	81, 80, 88, 89
# 7	83, 82, 86, 87	# 17	83, 82, 86, 87
# 8	85, 84, 84, 85	# 18	85, 84, 84, 85
# 9	87, 86, 82, 83	# 19	87, 86, 82, 83
# 10	89, 88, 80, 81	# 20	89, 88, 80, 81

Lesson 11

	a b c d		a b c d
# 1	90, 91, 99, 98	# 11	90, 91, 99, 98
# 2	92, 93, 97, 96	# 12	92, 93, 97, 96
# 3	94, 95, 95, 94	# 13	94, 95, 95, 94
# 4	96, 97, 93, 92	# 14	96, 97, 93, 92
# 5	98, 99, 91, 90	# 15	98, 99, 91, 90
# 6	91, 90, 98, 99	# 16	91, 90, 98, 99
# 7	93, 92, 96, 97	# 17	93, 92, 96, 97
# 8	95, 94, 94, 95	# 18	95, 94, 94, 95
# 9	97, 96, 92, 93	# 19	97, 96, 92, 93
# 10	99, 98, 90, 91	# 20	99, 98, 90, 91

Lesson 12

Review 1 A

# 1	98	# 6	97	# 11	59	# 16	56
# 2	87	# 7	86	# 12	68	# 17	67
# 3	76	# 8	75	# 13	75	# 18	78
# 4	65	# 9	64	# 14	86	# 19	89
# 5	54	# 10	53	# 15	95	# 20	96

Review 1 B

# 1	79	# 6	78	# 11	59	# 16	56
# 2	86	# 7	85	# 12	69	# 17	67
# 3	76	# 8	75	# 13	75	# 18	78
# 4	67	# 9	64	# 14	87	# 19	89
# 5	56	# 10	53	# 15	97	# 20	89

Summary 4 Answers Chapter 3

Lesson 13 Review 2 A

# 1	96	# 6	95	# 11	57	# 16	54
# 2	85	# 7	84	# 12	66	# 17	65
# 3	74	# 8	73	# 13	72	# 18	76
# 4	63	# 9	62	# 14	84	# 19	87
# 5	52	# 10	51	# 15	93	# 20	94

Review 2 B

# 1	77	# 6	76	# 11	57	# 16	54
# 2	84	# 7	83	# 12	67	# 17	65
# 3	74	# 8	73	# 13	73	# 18	76
# 4	65	# 9	62	# 14	85	# 19	87
# 5	54	# 10	51	# 15	95	# 20	86

Lesson 14 Review 1 A

	a	b	c
# 1	4	12	24
# 2	8	24	48
# 3	12	36	72
# 4	16	48	96
# 5	20	60	120
# 6	24	72	144
# 7	28	84	168
# 8	32	96	192
# 9	36	108	216
# 10	40	120	240

	a	b	c
# 11	15	30	49
# 12	30	60	98
# 13	45	90	147
# 14	60	120	196
# 15	75	150	245
# 16	90	180	294
# 17	105	210	343
# 18	120	240	392
# 19	135	270	441
# 20	150	300	490

Review 1 B

	a	b	c
# 1	2	3	4
# 2	4	6	8
# 3	6	9	12
# 4	8	12	16
# 5	10	15	20
# 6	12	18	24
# 7	14	21	28
# 8	16	24	32
# 9	18	27	36
# 10	20	30	40

	a	b	c
# 11	5	6	7
# 12	10	12	14
# 13	15	18	21
# 14	20	24	28
# 15	25	30	35
# 16	30	36	42
# 17	35	42	49
# 18	40	48	56
# 19	45	54	63
# 20	50	60	70

Summary 4 Answers Chapter 3

Lesson 15 Review 1 A

	a	b	c
# 1	2	3	4
# 2	4	6	8
# 3	6	9	12
# 4	8	12	16
# 5	10	15	20
# 6	12	18	24
# 7	14	21	28
# 8	16	24	32
# 9	18	27	36
# 10	20	30	40
# 11	5	6	7
# 12	10	12	14
# 13	15	18	21
# 14	20	24	28
# 15	25	30	35
# 16	30	36	42
# 17	35	42	49
# 18	40	48	56
# 19	45	54	63
# 20	50	60	70

Review 1 B

	a	b	c
# 1	1	1	1
# 2	2	2	2
# 3	3	3	3
# 4	4	4	4
# 5	5	5	5
# 6	6	6	6
# 7	7	7	7
# 8	8	8	8
# 9	9	9	9
# 10	10	10	10
# 11	1	1	1
# 12	2	2	2
# 13	3	3	3
# 14	4	4	4
# 15	5	5	5
# 16	6	6	6
# 17	7	7	7
# 18	8	8	8
# 19	9	9	9
# 20	10	10	10

Lesson 16 Review 1 A

	a	b	c	d
# 1	1	1	1	1
# 2	3	3	3	3
# 3	5	5	5	5
# 4	7	7	7	7
# 5	9	9	9	9
# 6	11	11	11	11
# 7	13	13	13	13
# 8	15	15	15	15
# 9	17	17	17	17
# 10	19	19	19	19

	a	b	c	d
# 11	1	1	1	1
# 12	3	3	3	3
# 13	5	5	5	5
# 14	7	7	7	7
# 15	9	9	9	9
# 16	11	11	11	11
# 17	13	13	13	13
# 18	15	15	15	15
# 19	17	17	17	17
# 20	19	19	19	19

Summary 4 Answers Chapter 3

Lesson 16 Review 1 B

	a	b	c	d		a	b	c	d
# 1	10	6	3	1	# 11	17	20	23	26
# 2	11	7	4	2	# 12	22	26	30	34
# 3	12	8	5	3	# 13	27	32	33	42
# 4	13	9	6	4	# 14	32	38	44	50
# 5	14	10	7	5	# 15	37	44	51	58
# 6	15	11	8	6	# 16	42	50	58	66
# 7	16	12	9	7	# 17	47	56	65	74
# 8	17	13	10	8	# 18	52	62	72	82
# 9	18	14	11	9	# 19	57	68	79	90
# 10	19	15	12	10	# 20	62	74	86	98

Lesson 17 Review 2 A

# 1	126	# 6	712	# 11	162	# 16	711
# 2	27	# 7	100	# 12	72	# 17	102
# 3	24	# 8	94	# 13	45	# 18	99
# 4	13	# 9	777	# 14	23	# 19	480
# 5	149	# 10	810	# 15	146	# 20	800

Review 2 B

# 1	180	# 6	700	# 11	180	# 16	500
# 2	24	# 7	60	# 12	35	# 17	100
# 3	21	# 8	100	# 13	12	# 18	40
# 4	4	# 9	300	# 14	54	# 19	300
# 5	160	# 10	300	# 15	150	# 20	400

Lesson 18 Test 3 A

# 1	162	# 6	714	# 11	216	# 16	713
# 2	18	# 7	95	# 12	24	# 17	96
# 3	30	# 8	99	# 13	28	# 18	94
# 4	13	# 9	703	# 14	23	# 19	465
# 5	181	# 10	800	# 15	174	# 20	800

Test 3 B

# 1	252	# 6	715	# 11	270	# 16	714
# 2	28	# 7	97	# 12	30	# 17	98
# 3	24	# 8	114	# 13	18	# 18	94
# 4	31	# 9	250	# 14	47	# 19	330
# 5	176	# 10	750	# 15	179	# 20	850

Summary 4 Answers Chapter 3

Lesson 19 Review 4 A

# 1	240	# 6	400
# 2	27	# 7	200
# 3	28	# 8	60
# 4	20	# 9	444
# 5	190	# 10	300

Review 4 B

# 1	600	# 6	700
# 2	28	# 7	100
# 3	32	# 8	60
# 4	15	# 9	555
# 5	140	# 10	500

Lesson 20 Final Exam 4 A

# 1	240	# 6	600
# 2	24	# 7	100
# 3	28	# 8	60
# 4	20	# 9	333
# 5	170	# 10	200

Final Exam 4 B

# 1	480	# 6	700
# 2	30	# 7	104
# 3	18	# 8	60
# 4	45	# 9	250
# 5	160	# 10	200

Math Help

for College

Summary 5

Summary 5 Pretest 5

Date **Name**

 # 1 Subtract

$$\begin{array}{r} 4000 \\ \underline{-\ 2182} \end{array}$$

 # 2 Multiply

$$\begin{array}{r} 678 \\ \underline{\times\ 7} \end{array}$$

 # 3

$$\begin{array}{r} 83 \\ \underline{\times\ 75} \end{array}$$

 # 4 Divide $6030 \div 9$

 # 5 $7332 \div 78$

Evaluate

 # 6 $34 \times 63 \div 7$

 # 7 $1664 \div 8 \times 9$

 # 8 $3312 \div 9 \div 8$

 # 9 $1000 - 75 \times 9$

 # 10 $6304 - 7101 \div 9$

Summary 5

Chapter 1

Four-Digit

Numbers 2

Summary 5 Chapter 1 Contents
Four-Digit Numbers
Lesson 1 - 10
Basic Problems

Work with lesson 1, read rules and examples, 2) Solve the problems from contents, 3) Check answers in the end of chapter 1, 4) Work with the next lessons 2, 3, 4, 5, ..., 5) Solve Test 1 A, 6) Grading Scale: A =10 points, B=9points, C=8 points, D=7 points, F=0-6 points, 7) If you solve 10 problems correctly, you get an A, 8) If have an A, you can move on to next chapter, 9) If your grade is lower than an A, work with the mistakes, then solve Test 1B, 10) When you finish all of the chapters, solve Final Exam in last chapter.

Lesson 1 Three-digit numbers.
Solve the problems # 2, # 4. Pretest # 1 - # 10
Lesson 2 Four-digit numbers. Solve the problems # 2, # 4

Lesson 3 Table of Multiplication. Solve the problems # 1- # 10
Lesson 4 Multiplication numbers.
Solve the problems Review 1 A # 4 (a, b, c, d, e)
Review 1 B # 8 (a, b, c, d, e)

Lesson 5 Division numbers.
Solve the problems Review 1 A # 4 (a, b, c, d,)
Review 1 B # 8 (a, b, c, d)
Lesson 6 Four Operations with numbers.
Solve Review 1 A # 5, # 10, Review 1 B # 5, # 10

Lesson 7 Solve Review 2 A, Review 2 B # 1, 3, 5, 7, 9, 11, 13, 15, 17, 19
Lesson 8 Signs for operations. Solve the problems
Review 1 A # 5, # 10, # 15, # 20, Review 1B # 5, # 10, # 15, # 20

Lesson 9 Solve Review 1 A # 1 - # 10 all, Solve Review 1 B # 1 - # 10 all
Lesson 10 Test 1 A, Test 1 B Test for 50 minutes (10 problems)
Solve Test 1 A # 1 - # 10 all, Solve Test 1 B # 1 - # 10 all

Summary 5 Chapter 1 Lesson 1

Three-digit numbers

Three-digit numbers content three digits.
This is numbers 100, 101, 102 …999.
Least three-digit number: 100. Greatest three-digit number: 999
Digits for three-digit numbers from right to the left:
Ones, tens, and hundreds 468 = 400 + 60 + 8
One-digit numbers less than two or three-digit numbers
Three-digit numbers greater than one or two-digit numbers
To compare numbers, compare first hundred, after that tens, then ones

$423 < 456$ $\qquad$ $789 > 786$

Two numbers can add in any order. Sum will be the same.

200+300 = 500 $\qquad$ 300+200=500

The sum of a number and zero is the same number

500+0 = 500 $\qquad$ 0+500=500

Difference any number and zero is the same number 500 – 0 = 500
Difference two the same number is zero 500 – 500 = 0

Adding three-digit numbers:

Add the ones. Add the tens. Add the hundreds.
Subtracting three-digit numbers:
Subtract the ones. Subtract the tens. Subtract the hundreds.

		1 1	71414
246	678	154	8 5 4
+ 351	- 432	+789	- 2 8 6
597	246	943	5 6 8

Add or subtract three-digit numbers

	a	b	c	d	e
# 1	273 +598	286 +589	299 +592	377 +559	429 +546
# 2	294 +532	378 +476	392 +469	322 +448	336 +439
# 3	966 -357	945 -378	924 -399	903 -504	882 -525
# 4	989 -598	943 -575	827 -499	862 -483	815 -437

Summary 5 Chapter 1 Lesson 2
Four-digit numbers

Four-digit numbers content four digits: 1000, 100, …9999
Least four-digit number: 1000. Greatest four-digit number: 9999
First period - Ones. First period has three digits. 234
Digits count from right to the left: ones, tens, hundreds, thousands
2468 = 2000 + 400 + 60 + 8
Four-digit numbers greater than one, two, three-digit numbers
One-digit numbers always less than two, three, four-digit numbers
To compare numbers, compare thousands, hundreds, tens, and ones:

1234 < 1239 1238 > 1230 1234=1234

Two numbers can add in any order. The sum will be the same.

2000 + 3000 = 5000 3000 + 2000 = 5000

The sum of a number and zero is the same number

5000 + 0 = 5000 0 + 5000 = 5000

Difference any number and zero is the same number: 5000–0= 5000
The difference two the same number is zero: 5000 – 5000 = 0
Adding four-digit numbers:
Add the ones. Add the tens. Add the hundreds. Add the thousands.
Subtracting four digits numbers: Subtract the ones. Subtract the tens. Subtract the hundreds. Subtract the thousands.

		1	8 13 14 16
3246	9678	5678	9 4 5 6
+ 4351	- 5432	+1050	- 3 9 8 7
7597	4246	6728	5 4 6 9

Add or subtract four-digit numbers

	a	b	c	d	e
# 1	3939	4848	5757	6969	7979
	+ 1393	+ 1484	+ 1575	+ 1696	+ 1797
# 2	4949	5858	6767	7878	8989
	+ 1494	+ 1585	+ 1676	+ 1787	+ 898
# 3	3110	4110	5110	6110	7554
	- 1919	- 2828	- 3737	- 4646	- 5959
# 4	4221	5221	6221	7221	8554
	- 2929	- 3838	- 4747	- 5656	- 6868

Summary 5 Chapter 1 Lesson 3

Multiplication numbers

Third operation with numbers is multiplication × or ·
Multiplication is the operation of repeated addition
$2\times 3 = 2 + 2 + 2 = 6$ or $2 \cdot 3 = 2 + 2 + 2 = 6$
First number in multiplication called the factor 2
Second number in multiplication called the factor 3
Result of multiplication called the product 6

Properties of multiplication

Two numbers can multiply in any order. The product will be the same
$2 \times 3 = 2 + 2 + 2 = 6$ $3 \times 2 = 3 + 3 = 6$
The product of a number and zero is zero
$5 \times 0 = 0$ $0 \times 5 = 0$
The product of a number and one is the same number
$5 \times 1 = 5$ $1 \times 5 = 5$

Division numbers

Division means separated number by some equal groups $6\div2=3$
First number in division is called dividend 6
Second number in division is called divisor 2
Result of division is called quotient 3
Exact Division Remainder $R = 0$ $6\div2=3$
Any number divided by one is the same number $5 \div 1 = 5$
Any number divided by the itself is one $5 \div 5 = 1$
Zero divided by any number is zero $0 \div 5 = 0$
Division by zero is not allowed $5 \div 0$ No number
To check multiplication use division
$2\times3=6$ Check $6\div2=3$ or $6\div3=2$
To check division use multiplication and division
$6\div2=3$ Check $2\times3=6$ or $6\div3=2$

Summary 5 Chapter 1 Lesson 3
Memorize the Table of Multiplication

Number 5 from the left column multiply by number 4 from first row to get answer 20. $5 \times 4 = 20$ $20 \div 4 = 5$ $20 \div 5 = 4$

×	2	3	4	5	6	7	8	9
1	2	3	4	5	6	7	8	9
2	4	6	8	10	12	14	16	18
3	6	9	12	15	18	21	24	27
4	8	12	16	20	24	28	32	36
5	10	15	20	25	30	35	40	45
6	12	18	24	30	36	42	48	54
7	14	21	28	35	42	49	56	63
8	16	24	32	40	48	56	64	72
9	18	27	36	45	54	63	72	81
10	20	30	40	50	60	70	80	90

Complete the Table of Division by using Table of Multiplication

	a	b	c	d	e	f	g	h	i
	×	2		4		6		8	
#1	1								9
#2									18
#3	3								
#4								32	
#5	5						35		
#6						36			
#7	7				35				
#8				32					
#9	9		27						
#10		20							

Summary 5 Chapter 1 Lesson 4

Multiplication Numbers

Multiply numbers in row

1) 96×7
2) $96 = 90 + 6$
3) $90 \times 7 = 630$
4) $6 \times 7 = 42$
5) $630 + 42 = 672$
6) $96 \times 7 = 672$

Review 1 A

Multiply numbers

	a	b	c	d	e
# 1	90×2	91×2	92×2	93×2	94×2
# 2	90×3	91×3	92×3	93×3	94×3
# 3	90×4	91×4	92×4	93×4	94×4
# 4	90×5	91×5	92×5	93×5	94×5
# 5	90×6	91×6	92×6	93×6	94×6

Multiply numbers in column

$$\begin{array}{r} {}^{4} \\ 96 \\ \times \underline{7} \\ 672 \end{array}$$

1) $7 \times 6 = 42$
2) Put 2 below 7
3) Put 4 above 9
4) $7 \times 9 = 63$
5) $63 + 4 = 67$
6) Put 67 ahead 2

Review 1 B

Multiply numbers

	a	b	c	d	e
# 6	$95 \times \underline{7}$	$96 \times \underline{7}$	$97 \times \underline{7}$	$98 \times \underline{7}$	$99 \times \underline{7}$
# 7	$95 \times \underline{8}$	$96 \times \underline{8}$	$97 \times \underline{8}$	$98 \times \underline{8}$	$99 \times \underline{8}$
# 8	$95 \times \underline{9}$	$96 \times \underline{9}$	$97 \times \underline{9}$	$98 \times \underline{9}$	$99 \times \underline{9}$

Summary 5 Chapter 1 Lesson 5

Divide numbers in column

$5 \times 1 = 5$ $5 \times 2 = 10$ $5 \times 3 = 15$ $5 \times 4 = 20$

Division from right to left or Division from left to right

29	1) $5 \times 2 = 10$	145 | 5
5 | 145	2) $14 - 10 = 4$	-10 29
- 10	3) Bring down the 5	45
45	4) $5 \times 9 = 45$	-45
- 45	5) $45 - 45 = 0$	0
0	6) R = 0	

Review 1 A

Divide numbers

	a	b	c	d
# 1	$40 \div 2$	$63 \div 3$	$116 \div 4$	$140 \div 5$
# 2	$44 \div 2$	$69 \div 3$	$108 \div 4$	$130 \div 5$
# 3	$48 \div 2$	$75 \div 3$	$100 \div 4$	$120 \div 5$
# 4	$52 \div 2$	$81 \div 3$	$92 \div 4$	$110 \div 5$
# 5	$56 \div 2$	$87 \div 3$	$84 \div 4$	$100 \div 5$

Divide numbers in column

$9 \times 1 = 9$ $9 \times 2 = 18$ $9 \times 3 = 27$

Division from right to left or Division from left to right

29	1) $9 \times 2 = 18$	261 | 9
9 | 261	2) $26 - 18 = 8$	-18 29
- 18	3) Bring down the 1	81
81	4) $9 \times 9 = 81$	-81
- 81	5) $81 - 81 = 0$	0
0	6) R = 0	

Review 1 B

Divide numbers

	a	b	c	d
# 6	$150 \div 6$	$168 \div 7$	$192 \div 8$	$225 \div 9$
# 7	$162 \div 6$	$182 \div 7$	$176 \div 8$	$207 \div 9$
# 8	$174 \div 6$	$196 \div 7$	$160 \div 8$	$189 \div 9$

Summary 5 Chapter 1 Lesson 6
Four Operations with numbers
Multiplication numbers

To multiply numbers, work in any order

$9 \times 8 \times 7 = 72 \times 7 = 504$ $9 \times 8 \times 7 = 9 \times 56 = 504$

Multiplication and Division numbers

To multiply and divide numbers work from left to right

$9 \times 6 \div 3 = 54 \div 3 = 18$ $4 \times 81 \div 9 = 324 \div 9 = 36$

Division and Multiplication numbers

To divide and multiply numbers work from left to right

$24 \div 6 \times 2 = 4 \times 2 = 8$ $81 \div 9 \times 4 = 9 \times 4 = 36$

Division numbers

To divide numbers, work from left to right

$24 \div 4 \div 2 = 6 \div 2 = 3$ $810 \div 9 \div 9 = 90 \div 9 = 10$

Multiplication and Addition numbers

To multiply and add numbers always first multiply than add

$9 \times 6 + 9 = 54 + 9 = 63$ $9 + 9 \times 9 = 9 + 81 = 90$

Multiplication and Subtraction numbers

To multiply and subtract numbers always first multiply than subtract

$9 \times 9 - 9 = 81 - 9 = 72$ $90 - 9 \times 9 = 90 - 81 = 9$

Division and Addition numbers

To divide and add numbers always first divide than add

$50 \div 5 + 39 = 10 + 39 = 49$ $88 + 171 \div 9 = 88 + 19 = 107$

Division and Subtraction numbers

To divide and subtract numbers always first divide than subtract

$108 \div 6 - 9 = 18 - 9 = 9$ $500 - 126 \div 7 = 500 - 18 = 482$

Summary 5 Chapter 1 Lesson 6
Four Operations with numbers
Review 1 A
Evaluate

# 1	$9 \times 2 \times 9$	# 11	$9 \times 3 \times 8$
# 2	$9 \times 18 \div 9$	# 12	$8 \times 27 \div 9$
# 3	$54 \div 9 \times 5$	# 13	$63 \div 9 \times 4$
# 4	$819 \div 9 \div 7$	# 14	$828 \div 9 \div 4$
# 5	$100 + 9 \times 9$	# 15	$110 + 8 \times 8$
# 6	$80 \times 6 - 90$	# 16	$90 \times 7 - 80$
# 7	$140 - 9 \times 5$	# 17	$150 - 9 \times 6$
# 8	$840 \div 6 - 80$	# 18	$900 \div 6 - 90$
# 9	$200 + 207 \div 9$	# 19	$210 + 200 \div 8$
# 10	$500 - 228 \div 6$	# 20	$510 - 322 \div 7$

Review 1 B

Evaluate

# 1	$9 \times 4 \times 7$	# 11	$9 \times 5 \times 6$
# 2	$7 \times 36 \div 9$	# 12	$6 \times 45 \div 9$
# 3	$72 \div 9 \times 3$	# 13	$81 \div 9 \times 2$
# 4	$837 \div 9 \div 3$	# 14	$846 \div 9 \div 2$
# 5	$120 + 8 \times 7$	# 15	$130 + 7 \times 7$
# 6	$90 \times 6 - 89$	# 16	$80 \times 7 - 99$
# 7	$160 - 9 \times 7$	# 17	$170 - 9 \times 8$
# 8	$960 \div 6 - 99$	# 18	$873 \div 9 - 69$
# 9	$220 + 189 \div 7$	# 19	$230 + 174 \div 6$
# 10	$520 - 432 \div 8$	# 20	$530 - 558 \div 9$

Summary 5 Chapter 1 Lesson 7
Four operations with numbers
Review 2 A
Perform operations with numbers

# 1	$98\times 4\times 5$	# 11	$96 \times 8 \times 5$
# 2	$96\times 40 \div 8$	# 12	$98\times 45 \div 9$
# 3	$35 \div 7\times 986$	# 13	$30 \div 6\times 978$
# 4	$7848 \div 9 \div 8$	# 14	$6048 \div 8 \div 7$
# 5	$985\times 8 + 120$	# 15	$120 + 976\times 5$
# 6	$95\times 98 - 10$	# 16	$96\times 95 - 20$
# 7	$8200 - 85\times 96$	# 17	$8000 - 84\times 95$
# 8	$7264\div 8 + 92$	# 18	$5436\div 6 + 94$
# 9	$8544\div 96 - 19$	# 19	$8526\div 98 - 27$
# 10	$9598 - 9310\div 95$	# 20	$9396 - 9120\div 95$

Review 2 B
Perform operations with numbers

# 1	$98\times 5\times 4$	# 11	$96\times 6\times 5$
# 2	$96\times 48\div 8$	# 12	$98\times 48 \div 6$
# 3	$30 \div 6\times 908$	# 13	$40 \div 8\times 906$
# 4	$5136 \div 8 \div 6$	# 14	$3240 \div 6 \div 5$
# 5	$975\times 8 + 200$	# 15	$160 + 968\times 5$
# 6	$95\times 94 - 30$	# 16	$95\times 96 - 20$
# 7	$8200 - 86\times 95$	# 17	$8000- 84\times 95$
# 8	$8172\div 9+92$	# 18	$94 + 7248\div 8$
# 9	$8633\div 97 - 19$	# 19	$8352\div 96 - 27$
# 10	$9895 - 9310\div 98$	# 20	$9695 - 9120\div 96$

Summary 5 Chapter 1 Lesson 8
Signs for operations
Review 1 A

Put + or – between numbers that each problem was correct

20 ? 60 = 80   20 + 60 = 80   60 ? 20 = 40   60 - 20 = 40

	a	b	c	d
# 1	20 ? 40=60	30 ? 30=60	40 ? 20=60	50 ? 10=60
# 2	60 ? 10=50	70 ? 10=60	80 ? 10=70	90 ? 10=80
# 3	21 ? 42=63	31 ? 32=63	41 ? 22=63	51 ? 12=63
# 4	61 ? 12=49	71 ? 12=59	81 ? 12=69	91 ? 12=79
# 5	22 ? 44=66	32 ? 34=66	42 ? 24=66	52 ? 14=66
# 6	62 ? 14=48	72 ? 14=58	82 ? 14=68	92 ? 14=78
# 7	23 ? 46=69	33 ? 36=69	43 ? 26=69	53 ? 16=69
# 8	63 ? 16=47	73 ? 16=57	83 ? 16=67	93 ? 16=77
# 9	24 ? 48=72	34 ? 38=72	44 ? 28=72	54 ? 18=72
# 10	64 ? 18=46	74 ? 18=56	84 ? 18=66	94 ? 18=76
# 11	25 ? 41=66	35 ? 31=66	45 ? 21=66	55 ? 11=66
# 12	65 ? 11=54	75 ? 11=64	85 ? 11=74	95 ? 11=84
# 13	26 ? 43=69	36 ? 33=69	46 ? 23=69	56 ? 13=69
# 14	66 ? 13=53	76 ? 13=63	86 ? 13=73	96 ? 13=83
# 15	27 ? 45=72	37 ? 35=72	47 ? 25=72	57 ? 15=72
# 16	67 ? 15=52	77 ? 15=62	87 ? 15=72	97 ? 15=82
# 17	28 ? 47=75	38 ? 37=75	48 ? 27=75	58 ? 17=75
# 18	68 ? 17=51	78 ? 17=61	88 ? 17=71	98 ? 17=81
# 19	29 ? 49=78	39 ? 39=78	49 ? 29=78	59 ? 19=78
# 20	69 ? 19=50	79 ? 19=60	89 ? 19=70	99 ? 19=80

Summary 5 Chapter 1 Lesson 8
Signs for operations
Review 1 B

Put × or ÷ between numbers that each problem was correct

6 ? 2 = 12 6 × 2 = 12 6 ? 2 = 3 6 ÷ 2 = 3

	a	b	c	d
# 1	6 ? 0 = 0	7 ? 0 = 0	8 ? 0 = 0	9 ? 0 = 0
# 2	6 ? 6 = 1	7 ? 7 = 1	8 ? 8 = 1	9 ? 9 = 1
# 3	6 ? 2 = 12	7 ? 2 = 14	8 ? 2 = 16	9 ? 2 = 18
# 4	12 ? 6=2	14 ? 7 = 2	16 ? 8 = 2	18 ? 9 = 2
# 5	6 ? 3 = 18	7 ? 3 = 21	8 ? 3 = 24	9 ? 3 = 27
# 6	18 ? 6 = 3	21 ? 7 = 3	24 ? 8 = 3	27 ? 9 = 3
# 7	6 ? 4 = 24	7 ? 4 = 28	8 ? 4 = 32	9 ? 4 = 36
# 8	24 ? 6 = 4	28 ? 7 = 4	32 ? 8 = 4	36 ? 9 = 4
# 9	6 ? 5 = 30	7 ? 5 = 35	8 ? 5 = 40	9 ? 5 = 45
# 10	30 ? 6 = 5	35 ? 7 = 5	40 ? 8 = 5	45 ? 9 = 5
# 11	6 ? 6 = 36	7 ? 6 = 42	8 ? 6 = 48	9 ? 6 = 54
# 12	36 ? 6 = 6	42 ? 7 = 6	48 ? 8 = 6	54 ? 9 = 6
# 13	6 ? 7 = 42	7 ? 7 = 49	8 ? 7 = 56	9 ? 7 = 63
# 14	42 ? 6 = 7	49 ? 7 = 7	56 ? 8 = 7	63 ? 9 = 7
# 15	6 ? 8 = 48	7 ? 8 = 56	8 ? 8 = 64	9 ? 8 = 72
# 16	48 ? 6 = 8	56 ? 7 = 8	64 ? 8 = 8	72 ? 9 = 8
# 17	6 ? 9 = 54	7 ? 9 = 63	8 ? 9 = 72	9 ? 9 = 81
# 18	54 ? 6 = 9	63 ? 7 = 9	72 ? 8 = 9	81 ? 9 = 9
# 19	6 ? 10 = 60	7 ? 10 = 70	8 ? 10 = 80	9 ? 10 = 90
# 20	60 ? 6 = 10	70 ? 7 = 10	80 ? 8 = 10	90 ? 9 = 10

Summary 5 Chapter 1 Lesson 9
Review 1 A for Test 1
Evaluate

1 $9 \times 5 \times 8$

2 $8 \times 36 \div 9$

3 $72 \div 9 \times 4$

4 $756 \div 9 \div 4$

5 $110 + 9 \times 8$

6 $80 \times 7 - 90$

7 $150 - 9 \times 7$

8 $960 \div 6 - 80$

9 $240 + 400 \div 8$

10 $450 - 360 \div 9$

Review 1 B for Test 1
Evaluate

1 $9 \times 4 \times 6$

2 $6 \times 54 \div 9$

3 $72 \div 9 \times 2$

4 $756 \div 9 \div 2$

5 $130 + 8 \times 7$

6 $80 \times 9 - 90$

7 $170 - 9 \times 8$

8 $588 \div 6 - 29$

9 $210 + 150 \div 6$

10 $720 - 540 \div 9$

Summary 5 Chapter 1 Lesson 10
Four operations with numbers
Test 1 A
Evaluate

1 $9 \times 5 \times 6$

2 $9 \times 32 \div 8$

3 $81 \div 9 \times 6$

4 $640 \div 8 \div 4$

5 $220 + 8 \times 7$

6 $70 \times 9 - 80$

7 $170 - 8 \times 9$

8 $980 \div 7 - 90$

9 $225 + 600 \div 8$

10 $450 - 400 \div 8$

Summary 5 Chapter 1 Lesson 10
Four operations with numbers
Test 1 B
Evaluate

 # 1 $9 \times 5 \times 4$

 # 2 $7 \times 36 \div 9$

 # 3 $63 \div 9 \times 5$

 # 4 $630 \div 7 \div 2$

 # 5 $330 + 7 \times 9$

 # 6 $60 \times 8 - 80$

 # 7 $180 - 7 \times 9$

 # 8 $990 \div 9 - 20$

 # 9 $330 + 560 \div 8$

 # 10 $400 - 450 \div 9$

Summary 5 Answers Chapter 1

Pretest 1) 1818 2) 4746 3) 6225 4) 670 5) 94
6) 306 7) 1872 8) 46 9) 325 10) 5515

Lesson 1

	a, b, c, d, e		a, b, c, d, e
# 1	871, 875, 891, 936, 975	# 3	609, 567, 525, 399, 357
# 2	826, 854, 861, 770, 775	# 4	391, 368, 328, 379, 378

Lesson 2

	a	b	c	d	e
#1	5332	6332	7332	8665	9776
#2	6443	7443	8443	9665	9887
#3	1191	1282	1373	1464	1595
#4	1292	1383	1474	1565	1686

Lesson 3 Table of division

	a	b	c	d	e	f	g	h	i
	×	2	3	4	5	6	7	8	9
# 1	1	2	3	4	5	6	7	8	9
# 2	2	4	6	8	10	12	14	16	18
# 3	3	6	9	12	15	18	21	24	27
# 4	4	8	12	16	20	24	28	32	36
# 5	5	10	15	20	25	30	35	40	45
# 6	6	12	18	24	30	36	42	48	54
# 7	7	14	21	28	35	42	49	56	63
# 8	8	16	24	32	40	48	56	64	72
# 9	9	18	27	36	45	54	63	72	81
# 10	10	20	30	40	50	60	70	80	90

Summary 5 Answers Chapter 1

Lesson 4

	a, b, c, d, e		a, b, c, d, e
# 1	180, 182, 184, 186, 188	# 5	540, 546, 552, 558, 564
# 2	270, 273, 276, 279, 282	# 6	665, 672, 679, 686, 693
# 3	360, 364, 368, 372, 376	# 7	760, 768, 776, 784, 792
# 4	450, 455, 460, 465, 470	# 8	855, 864, 873, 882, 891

Lesson 5

	a, b, c, d		a, b, c, d
# 1	20, 21, 29, 28	# 5	28, 29, 21, 20
# 2	22, 23, 27, 26	# 6	25, 24, 24, 25
# 3	24, 25, 25, 24	# 7	27, 26, 22, 23
# 4	26, 27, 23, 22	# 8	29, 28, 20, 21

Lesson 6

Review 1 A

# 1	162	# 2	18	# 3	30	#4	13	#5	181
# 6	390	# 7	95	# 8	60	# 9	223	#10	462
# 11	216	#12	24	#13	28	#14	23	#15	174
#16	550	#17	96	#18	60	#19	235	#20	464

Review 1 B

#1	252	#2	28	#3	24	#4	31	#5	176
#6	451	#7	97	#8	61	#9	247	#10	466
#11	270	#12	30	#13	18	#14	47	#15	179
#16	461	#17	98	#18	28	#19	259	#20	468

Lesson 7 Review 2 A

#1	1960	#2	480	#3	4930	#4	109	#5	8000
#6	9300	#7	40	#8	1000	#9	70	#10	9500
#11	3840	#12	490	#13	4890	#14	108	#15	5000
#16	9100	#17	20	#18	1000	#19	60	#20	9300

Review 2 B

#1	1960	#2	576	#3	4540	#4	107	#5	8000
#6	8900	#7	30	#8	1000	#9	70	#10	9800
#11	2880	#12	784	#13	4530	#14	108	#15	5000
#16	9100	#17	20	#18	1000	#19	60	#20	9600

Summary 5 Answers Chapter 1

Lesson 8 Review 1 A

	a	b	c	d
# 1	+	+	+	+
# 2	-	-	-	-
# 3	+	+	+	+
# 4	-	-	-	-
# 5	+	+	+	+
# 6	-	-	-	-
# 7	+	+	+	+
# 8	-	-	-	-
# 9	+	+	+	+
#10	-	-	-	-

	a	b	c	d
#11	+	+	+	+
#12	-	-	-	-
#13	+	+	+	+
#14	-	-	-	-
#15	+	+	+	+
#16	-	-	-	-
#17	+	+	+	+
#18	-	-	-	-
#19	+	+	+	+
#20	-	-	-	-

Review 1 B

	a	b	c	d
#1	×	×	×	×
#2	÷	÷	÷	÷
#3	×	×	×	×
#4	÷	÷	÷	÷
#5	×	×	×	×
#6	÷	÷	÷	÷
#7	×	×	×	×
#8	÷	÷	÷	÷
#9	×	×	×	×
#10	÷	÷	÷	÷

	a	b	c	d
#11	×	×	×	×
#12	÷	÷	÷	÷
#13	×	×	×	×
#14	÷	÷	÷	÷
#15	×	×	×	×
#16	÷	÷	÷	÷
#17	×	×	×	×
#18	÷	÷	÷	÷
#19	×	×	×	×
#20	÷	÷	÷	÷

Summary 5 Answers Chapter 1

Lesson 9

Review 1 A

#1	360	#2	32	#3	32	#4	21	#5	182
#6	470	#7	87	#8	80	#9	290	#10	410

Review 1 B

#1	216	#2	36	#3	16	#4	42	#5	186
#6	630	#7	98	#8	69	#9	235	#10	660

Lesson 10

Test 1 A

#1	270	#2	36	#3	54	#4	20	#5	276
#6	550	#7	98	#8	50	#9	300	#10	400

Test 1 B

#1	180	#2	28	#3	35	#4	45	#5	393
#6	400	#7	117	#8	90	#9	400	#10	350

Summary 5

Chapter 2

Multiplication

Numbers 2

Summary 5 Chapter 2 Contents
Multiplication Numbers
Lesson 1 - 10
Basic Problems

Work with lesson 1, read rules and examples, 2) Solve the problems from contents, 3) Check answers in the end of chapter 1, 4) Work with the next lessons 2, 3, 4, 5, ..., 5) Solve Test 1 A, 6) Grading Scale: A =10 points, B=9points, C=8 points, D=7 points, F=0-6 points, 7) If you solve 10 problems correctly, you get an A, 8) If have an A, you can move on to next chapter, 9) If your grade is lower than an A, work with the mistakes, then solve Test 1B, 10) When you finish all of the chapters, solve Final Exam in last chapter.

Summary 5 Chapter 2 Lesson 1
Multiplication Numbers 1
Example 1

125	1) 125 = 100 + 20 + 5
×6	2) 6 × 5 = 30
30	3) 6 × 20 = 120
120	4) 6 × 100 = 600
+600	
750	

Multiply numbers

	a	b	c	d	e
# 1	120× 6	121× 6	122× 6	123× 6	124× 6
# 2	125× 6	126× 6	127× 6	128× 6	129× 6
# 3	130× 7	131× 7	132× 7	133× 7	134× 7
# 4	135× 7	136× 7	137× 7	138× 7	139× 7
# 5	140× 8	141× 8	142× 8	143× 8	144× 8
# 6	145× 8	146× 8	147× 8	148× 8	149× 8
# 7	150× 9	151× 9	152× 9	153× 9	154× 9
# 8	155× 9	156× 9	157× 9	158× 9	159× 9

Example 2

1 3	1) 6×5=30	6) Put 5 ahead 0
125	2) Put 0 below 6	7) Put 1 ahead 3
×6	3) Put 3 above 2	8) 6 × 1 =6
750	4) 6 × 2 = 12	9) 6 + 1 = 7
	5) 12 + 3 = 15	10) Put 7 ahead 5

Multiply numbers

	a	b	c	d	e
# 9	160× 6	161× 6	162× 6	163× 6	164× 6
# 10	165× 6	166× 6	167× 6	168×6	169× 6
# 11	170× 7	171× 7	172× 7	173× 7	174× 7
# 12	175× 7	176× 7	177× 7	178× 7	179× 7
# 13	180× 8	181× 8	182× 8	183× 8	184× 8
# 14	185× 8	186× 8	187× 8	188× 8	189× 8
# 15	190× 9	191× 9	192× 9	193× 9	194× 9
# 16	195× 9	196× 9	197× 9	198× 9	199× 9

Summary 5 Chapter 2 Lesson 2
Multiplication Numbers 2
Example 1

275	1) 275 = 200 + 70 + 5
×6	2) 6 × 5 = 30
30	3) 6 × 70 = 420
420	4) 6 × 200 = 1200
+ 1200	
1650	

Multiply numbers

	a	b	c	d	e
# 1	220× 6	221× 6	222× 6	223× 6	224× 6
# 2	225× 6	226× 6	227× 6	228× 6	229× 6
# 3	230× 7	231× 7	232× 7	233× 7	234× 7
# 4	235× 7	236× 7	237× 7	238× 7	239× 7
# 5	240× 8	241× 8	242× 8	243× 8	244× 8
# 6	245× 8	246× 8	247× 8	248× 8	249× 8
# 7	250× 9	251× 9	252× 9	253× 9	254× 9
# 8	255× 9	256× 9	257× 9	258× 9	259× 9

Example 2

4 3	1) 6×5 = 30	6) Put 5 ahead 0
275	2) Put 0 below 6	7) Put 4 ahead 3
×6	3) Put 3 above 7	8) 6 × 2 = 12
1650	4) 6 × 7 = 42	9) 12 + 4 = 16
	5) 42 + 3 = 45	10) Put 16 ahead 5

Multiply numbers

	a	b	c	d	e
# 9	260× 6	261× 6	262× 6	263× 6	264× 6
# 10	265× 6	266× 6	267× 6	268× 6	269× 6
# 11	270× 7	271× 7	272× 7	273× 7	274× 7
# 12	275× 7	276× 7	277× 7	278× 7	279× 7
# 13	280× 8	281× 8	282× 8	283× 8	284× 8
# 14	285× 8	286× 8	287× 8	288× 8	289× 8
# 15	290× 9	291× 9	292× 9	293× 9	294× 9
# 16	295× 9	296× 9	297× 9	298× 9	299× 9

Summary 5 Chapter 2 Lesson 3
Multiplication Numbers 3
Example 1

```
  325     1) 325 = 300 + 20 + 5
   ×6     2) 6 × 5 = 30
   30     3) 6 × 20 = 120
  120     4) 6 × 300 = 1800
+1800
 1950
```

Multiply numbers

	a	b	c	d	e
# 1	320× 6	321× 6	322× 6	323× 6	324× 6
# 2	325× 6	326× 6	327× 6	328× 6	329× 6
# 3	330× 7	331× 7	332× 7	333× 7	334× 7
# 4	335× 7	336× 7	337× 7	338× 7	339× 7
# 5	340× 8	341× 8	342× 8	343× 8	344× 8
# 6	345× 8	346× 8	347× 8	348× 8	349× 8
# 7	350× 9	351× 9	352× 9	353× 9	354× 9
# 8	355× 9	356× 9	357× 9	358× 9	359× 9

Example 2

```
 1 3     1) 6×5 = 30            6) Put 5 ahead 0
 325     2) Put 0 below 6       7) Put 1 ahead 3
  ×6     3) Put 3 above 2       8) 6 × 3 = 18
1950     4) 6 × 2 = 12          9) 18 + 1 = 19
         5) 12 + 3 = 15        10) Put 19 ahead 5
```

Multiply numbers

	a	b	c	d	e
# 9	360× 6	361× 6	362× 6	363× 6	364× 6
# 10	365× 6	366× 6	367× 6	368×6	369× 6
# 11	370× 7	371× 7	372× 7	373× 7	374× 7
# 12	375× 7	376× 7	377× 7	378× 7	379× 7
# 13	380× 8	381× 8	382× 8	383× 8	384× 8
# 14	385× 8	386× 8	387× 8	388× 8	389× 8
# 15	390× 9	391× 9	392× 9	393× 9	394× 9
# 16	395× 9	396× 9	397× 9	398× 9	399× 9

Summary 5 Chapter 2 Lesson 4
Multiplication Numbers 4
Example 1

```
  425     1) 425 = 400 + 20 + 5
   ×6     2) 6 × 5 = 30
   30     3) 6 × 20 = 120
  120     4) 6 × 400 = 2400
+2400
 2550
```

Multiply numbers

	a	b	c	d	e
# 1	420× 6	421× 6	422× 6	423× 6	424× 6
# 2	425× 6	426× 6	427× 6	428× 6	429× 6
# 3	430× 7	431× 7	432× 7	433× 7	434× 7
# 4	435× 7	436× 7	437× 7	438× 7	439× 7
# 5	440× 8	441× 8	442× 8	443× 8	444× 8
# 6	445× 8	446× 8	447× 8	448× 8	449× 8
# 7	450× 9	451× 9	452× 9	453× 9	454× 9
# 8	455× 9	456× 9	457× 9	458× 9	459× 9

Example 2

```
 1 3     1) 6×5 = 30            6) Put 5 ahead 0
 425     2) Put 0 below 6       7) Put 1 ahead 3
  ×6     3) Put 3 above 2       8) 6 × 4 = 24
2550     4) 6 × 2 = 12          9) 24 + 1 = 25
         5) 12 + 3 = 15        10) Put 25 ahead 5
```

Multiply numbers

	a	b	c	d	e
# 9	460× 6	461× 6	462× 6	463× 6	464× 6
# 10	465× 6	466× 6	467× 6	468×6	469× 6
# 11	470× 7	471× 7	472× 7	473× 7	474× 7
# 12	475× 7	476× 7	477× 7	478× 7	479× 7
# 13	480× 8	481× 8	482× 8	483× 8	484× 8
# 14	485× 8	486× 8	487× 8	488× 8	489× 8
# 15	490× 9	491× 9	492× 9	493× 9	494× 9
# 16	495× 9	496× 9	497× 9	498× 9	499× 9

Summary 5 Chapter 2 Lesson 5

Multiplication Numbers 5

Example 1

$$\begin{array}{r} 525 \\ \underline{\times 6} \\ 30 \\ 120 \\ \underline{+3000} \\ 3150 \end{array}$$

1) 525 = 500 + 20 + 5
2) 6 × 5 = 30
3) 6 × 20 = 120
4) 6 × 500 = 3000

Multiply numbers

	a	b	c	d	e
# 1	520× 6	521× 6	522× 6	523× 6	524× 6
# 2	525× 6	526× 6	527× 6	528× 6	529× 6
# 3	530× 7	531× 7	532× 7	533× 7	534× 7
# 4	535× 7	536× 7	537× 7	538× 7	539× 7
# 5	540× 8	541× 8	542× 8	543× 8	544× 8
# 6	545× 8	546× 8	547× 8	548× 8	549× 8
# 7	550× 9	551× 9	552× 9	553× 9	554× 9
# 8	555× 9	556× 9	557× 9	558× 9	559× 9

Example 2

$$\begin{array}{r} 1\,3 \\ 525 \\ \underline{\times 6} \\ 3150 \end{array}$$

1) 6×5 = 30
2) Put 0 below 6
3) Put 3 above 2
4) 6 × 2 = 12
5) 12 + 3 = 15
6) Put 5 ahead 0
7) Put 1 ahead 3
8) 6 × 5 = 30
9) 30 + 1 = 31
10) Put 31 ahead 5

Multiply numbers

	a	b	c	d	e
# 9	560× 6	561× 6	562× 6	563× 6	564× 6
# 10	565× 6	566× 6	567× 6	568× 6	569× 6
# 11	570× 7	571× 7	572× 7	573× 7	574× 7
# 12	575× 7	576× 7	577× 7	578× 7	579× 7
# 13	580× 8	581× 8	582× 8	583× 8	584× 8
# 14	585× 8	586× 8	587× 8	588× 8	589× 8
# 15	590× 9	591× 9	592× 9	593× 9	594× 9
# 16	595× 9	596× 9	597× 9	598× 9	599× 9

Summary 5 Chapter 2 Lesson 6
Multiplication Numbers 6
Example 1

625	1) 625 = 600 + 20 + 5
×6	2) 6 × 5 = 30
30	3) 6 × 20 = 120
120	4) 6 × 600 = 3600
+3600	
3750	

Multiply numbers

	a	b	c	d	e
# 1	620× 6	621× 6	622× 6	623× 6	624× 6
# 2	625× 6	626× 6	627× 6	628× 6	629× 6
# 3	630× 7	631× 7	632× 7	633× 7	634× 7
# 4	635× 7	636× 7	637× 7	638× 7	639× 7
# 5	640× 8	641× 8	642× 8	643× 8	644× 8
# 6	645× 8	646× 8	647× 8	648× 8	649× 8
# 7	650× 9	651× 9	652× 9	653× 9	654× 9
# 8	655× 9	656× 9	657× 9	658× 9	659× 9

Example 2

1 3	1) 6×5 = 30	6) Put 5 ahead 0
625	2) Put 0 below 6	7) Put 1 ahead 3
×6	3) Put 3 above 2	8) 6 × 6 = 36
3750	4) 6 × 2 = 12	9) 36 + 1 = 37
	5) 12 + 3 = 15	10) Put 37 ahead 5

Multiply numbers

	a	b	c	d	e
# 9	660× 6	661× 6	662× 6	663× 6	664× 6
# 10	665× 6	666× 6	667× 6	668× 6	669× 6
# 11	670× 7	671× 7	672× 7	673× 7	674× 7
# 12	675× 7	676× 7	677× 7	678× 7	679× 7
# 13	680× 8	681× 8	682× 8	683× 8	684× 8
# 14	685× 8	686× 8	687× 8	688× 8	689× 8
# 15	690× 9	691× 9	692× 9	693× 9	694× 9
# 16	695× 9	696× 9	697× 9	698× 9	699× 9

Summary 5 Chapter 2 Lesson 7
Multiplication Numbers 7
Example 1

725	1) 725 = 700 + 20 + 5
×6	2) 6 × 5 = 30
30	3) 6 × 20 = 120
120	4) 6 × 700 = 4200
+4200	
4350	

Multiply numbers

	a	b	c	d	e
# 1	720× 6	721× 6	722× 6	723× 6	724× 6
# 2	725× 6	726× 6	727× 6	728× 6	729× 6
# 3	730× 7	731× 7	732× 7	733× 7	734× 7
# 4	735× 7	736× 7	737× 7	738× 7	739× 7
# 5	740× 8	741× 8	742× 8	743× 8	744× 8
# 6	745× 8	746× 8	747× 8	748× 8	749× 8
# 7	750× 9	751× 9	752× 9	753× 9	754× 9
# 8	755× 9	756× 9	757× 9	758× 9	759× 9

Example 2

1 3	1) 6×5 = 30	6) Put 5 ahead 0
725	2) Put 0 below 6	7) Put 1 ahead 3
×6	3) Put 3 above 2	8) 6 × 7 = 42
4350	4) 6 × 2 = 12	9) 42 + 1 = 43
	5) 12 + 3 = 15	10) Put 43 ahead 5

Multiply numbers

	a	b	c	d	e
# 9	760× 6	761× 6	762× 6	763× 6	764× 6
# 10	765× 6	766× 6	767× 6	768×6	769× 6
# 11	770× 7	771× 7	772× 7	773× 7	774× 7
# 12	775× 7	776× 7	777× 7	778× 7	779× 7
# 13	780× 8	781× 8	782× 8	783× 8	784× 8
# 14	785× 8	786× 8	787× 8	788× 8	789× 8
# 15	790× 9	791× 9	792× 9	793× 9	794× 9
# 16	795× 9	796× 9	797× 9	798× 9	799× 9

Summary 5 Chapter 2 Lesson 8
Multiplication Numbers 8
Example 1

825	1) 825 = 800 + 20 + 5
×6	2) 6 × 5 = 30
30	3) 6 × 20 = 120
120	4) 6 × 800 = 4800
+4800	
4950	

Multiply numbers

	a	b	c	d	e
# 1	820× 6	821× 6	822× 6	823× 6	824× 6
# 2	825× 6	826× 6	827× 6	828× 6	829× 6
# 3	830× 7	831× 7	832× 7	833× 7	834× 7
# 4	835× 7	836× 7	837× 7	838× 7	839× 7
# 5	840× 8	841× 8	842× 8	843× 8	844× 8
# 6	845× 8	846× 8	847× 8	848× 8	849× 8
# 7	850× 9	851× 9	852× 9	853× 9	854× 9
# 8	855× 9	856× 9	857× 9	858× 9	859× 9

Example 2

1 3	1) 6×5 = 30	6) Put 5 ahead 0
825	2) Put 0 below 6	7) Put 1 ahead 3
×6	3) Put 3 above 2	8) 6 × 8 = 48
4950	4) 6 × 2 = 12	9) 48 + 1 = 49
	5) 12 + 3 = 15	10) Put 49 ahead 5

Multiply numbers

	a	b	c	d	e
# 9	860× 6	861× 6	862× 6	863× 6	864× 6
# 10	865× 6	866× 6	867× 6	868×6	869× 6
# 11	870× 7	871× 7	872× 7	873× 7	874× 7
# 12	875× 7	876× 7	877× 7	878× 7	879× 7
# 13	880× 8	881× 8	882× 8	883× 8	884× 8
# 14	885× 8	886× 8	887× 8	888× 8	889× 8
# 15	890× 9	891× 9	892× 9	893× 9	894× 9
# 16	895× 9	896× 9	897× 9	898× 9	899× 9

Summary 5 Chapter 2 Lesson 9
Review for Test 2 A
Multiplication Numbers 9
Example 1

```
  925     1) 925 = 900 + 20 + 5
   ×6     2) 6 × 5 = 30
   30     3) 6 × 20 = 120
  120     4) 6 × 900 = 5400
+5400
 5550
```

Multiply numbers

	a	b	c	d	e
# 1	920× 6	921× 6	922× 6	923× 6	924× 6
# 2	925× 6	926× 6	927× 6	928× 6	929× 6
# 3	930× 7	931× 7	932× 7	933× 7	934× 7
# 4	935× 7	936× 7	937× 7	938× 7	939× 7
# 5	940× 8	941× 8	942× 8	943× 8	944× 8
# 6	945× 8	946× 8	947× 8	948× 8	949× 8
# 7	950× 9	951× 9	952× 9	953× 9	954× 9
# 8	955× 9	956× 9	957× 9	958× 9	959× 9

Summary 5 Chapter 2 Lesson 9
Review for Test 2 B
Multiplication Numbers 9
Example 2

1 3	1) 6×5 = 30	6) Put 5 ahead 0
925	2) Put 0 below 6	7) Put 1 ahead 3
×6	3) Put 3 above 2	8) 6 × 9 = 54
5550	4) 6 × 2 = 12	9) 54 + 1 = 55
	5) 12 + 3 = 15	10) Put 55 ahead 5

Multiply numbers

	a	b	c	d	e
# 9	960× 6	961× 6	962× 6	963× 6	964× 6
# 10	965× 6	966× 6	967× 6	968× 6	969× 6
# 11	970× 7	971× 7	972× 7	973× 7	974× 7
# 12	975× 7	976× 7	977× 7	978× 7	979× 7
# 13	980× 8	981× 8	982× 8	983× 8	984× 8
# 14	985× 8	986× 8	987× 8	988× 8	989× 8
# 15	990× 9	991× 9	992× 9	993× 9	994× 9
# 16	995× 9	996× 9	997× 9	998× 9	999× 9

Summary 5 Chapter 2 Lesson 10

Multiplication Numbers

Test 2 A

Multiply numbers

# 1 $\begin{array}{r} 534 \\ \times \underline{\ \ 5} \end{array}$	# 2 $\begin{array}{r} 567 \\ \times \underline{5} \end{array}$	# 3 $\begin{array}{r} 656 \\ \times \underline{\ \ 6} \end{array}$	# 4 $\begin{array}{r} 678 \\ \times \underline{6} \end{array}$
# 5 $\begin{array}{r} 745 \\ \times \underline{7} \end{array}$	# 6 $\begin{array}{r} 789 \\ \times \underline{7} \end{array}$	# 7 $\begin{array}{r} 856 \\ \times \underline{8} \end{array}$	# 8 $\begin{array}{r} 867 \\ \times \underline{\ \ 8} \end{array}$
# 9 $\begin{array}{r} 967 \\ \times \underline{\ \ 9} \end{array}$	# 10 $\begin{array}{r} 978 \\ \times \underline{9} \end{array}$	# 11 $\begin{array}{r} 543 \\ \times \underline{\ \ 5} \end{array}$	# 12 $\begin{array}{r} 576 \\ \times \underline{5} \end{array}$
# 13 $\begin{array}{r} 675 \\ \times \underline{\ \ 6} \end{array}$	# 14 $\begin{array}{r} 687 \\ \times \underline{6} \end{array}$	# 15 $\begin{array}{r} 754 \\ \times \underline{7} \end{array}$	# 16 $\begin{array}{r} 798 \\ \times \underline{7} \end{array}$
# 17 $\begin{array}{r} 865 \\ \times \underline{8} \end{array}$	# 18 $\begin{array}{r} 876 \\ \times \underline{\ \ 8} \end{array}$	# 19 $\begin{array}{r} 976 \\ \times \underline{\ \ 9} \end{array}$	# 20 $\begin{array}{r} 985 \\ \times \underline{9} \end{array}$

Summary 5 Chapter 2 Lesson 10

Multiplication numbers

Test 2 B

Multiply numbers

# 1	546 × 5	# 2	658 × 5	# 3	645 × 6	# 4	689 × 6
# 5	756 × 7	# 6	768 × 7	# 7	867 × 8	# 8	879 × 8
# 9	978 × 9	# 10	987 × 9	# 11	564 × 5	# 12	685 × 5
# 13	654 × 6	# 14	698 × 6	# 15	764 × 7	# 16	786 × 7
# 17	876 × 8	# 18	897 × 8	# 19	967 × 9	# 20	976 × 9

Summary 5 Answers Chapter 2

Lesson 1

	a	b	c	d	e		a	b	c	d	e
# 1	720	726	732	738	744	# 9	960	966	972	978	984
# 2	750	756	762	768	774	# 10	990	996	1002	1008	1014
# 3	910	917	924	931	938	# 11	1190	1197	1204	1211	1218
# 4	945	952	959	966	973	# 12	1225	1232	1239	1246	1253
# 5	1120	1128	1136	1144	1152	# 13	1440	1448	1456	1464	1472
# 6	1160	1168	1176	1184	1192	# 14	1480	1488	1496	1504	1512
# 7	1350	1359	1368	1377	1386	# 15	1710	1719	1728	1737	1746
# 8	1395	1404	1413	1422	1431	# 16	1755	1764	1773	1782	1791

Lesson 2

	a	b	c	d	e		a	b	c	d	e
# 1	1320	1326	1332	1338	1344	# 9	1560	1566	1572	1578	1584
# 2	1350	1356	1362	1368	1374	# 10	1590	1596	1602	1608	1614
# 3	1610	1617	1624	1631	1638	# 11	1890	1897	1904	1911	1918
# 4	1645	1652	1659	1666	1673	# 12	1925	1932	1939	1946	1953
# 5	1920	1928	1936	1944	1952	# 13	2240	2248	2256	2264	2272
# 6	1960	1968	1976	1984	1992	# 14	2280	2288	2296	2304	2312
# 7	2250	2259	2268	2277	2286	# 15	2610	2619	2628	2637	2646
# 8	2295	2304	2313	2322	2331	# 16	2655	2664	2673	2682	2691

Lesson 3

	a	b	c	d	e		a	b	c	d	e
# 1	1920	1926	1932	1938	1944	# 9	2160	2166	2172	2178	2184
# 2	1950	1956	1962	1968	1974	# 10	2190	2196	2202	2208	2214
# 3	2310	2317	2324	2331	2338	# 11	2590	2597	2604	2611	2618
# 4	2345	2352	2359	2366	2373	# 12	2625	2632	2639	2646	2653
# 5	2720	2728	2736	2744	2752	# 13	3040	3048	3056	3064	3072
# 6	2760	2768	2776	2784	2792	# 14	3080	3088	3096	3104	3112
# 7	3150	3159	3168	3177	3186	# 15	3510	3519	3528	3537	3546
# 8	3195	3204	3213	3222	3231	# 16	3555	3564	3573	3582	3591

Summary 5 Answers Chapter 2

Lesson 4

	a	b	c	d	e		a	b	c	d	e
# 1	2520	2526	2532	2538	2544	# 9	2760	2766	2772	2778	2784
# 2	2550	2556	2562	2568	2574	# 10	2790	2796	2802	2808	2814
# 3	3010	3017	3024	3031	3038	# 11	3290	3297	3304	3311	3318
# 4	3045	3052	3059	3066	3073	# 12	3325	3332	3339	3346	3353
# 5	3520	3528	3536	3544	3552	# 13	3840	3848	3856	3864	3872
# 6	3560	3568	3576	3584	3592	# 14	3880	3888	3896	3904	3912
# 7	4050	4059	4068	4077	4086	# 15	4410	4419	4428	4437	4446
# 8	4095	4104	4113	4122	4131	# 16	4455	4464	4473	4482	4491

Lesson 5

	a	b	c	d	e		a	b	c	d	e
# 1	3120	3126	3132	3138	3144	#9	3360	3366	3372	3378	3384
# 2	3150	3156	3162	3168	3174	#10	3390	3396	3402	3408	3414
# 3	3710	3717	3724	3731	3738	#11	3990	3997	4004	4011	4018
# 4	3745	3752	3759	3766	3773	#12	4025	4032	4039	4046	4053
# 5	4320	4328	4336	4344	4352	#13	4640	4648	4656	4664	4672
# 6	4360	4368	4376	4384	4392	#14	4680	4688	4696	4704	4712
# 7	4950	4959	4968	4977	4986	#15	5310	5319	5328	5337	5346
# 8	4995	5004	5013	5022	5031	#16	5355	5364	5373	5382	5391

Lesson 6

	a	b	c	d	e		a	b	c	d	e
# 1	3720	3726	3732	3738	3744	# 9	3960	3966	3972	3978	3984
# 2	3750	3756	3762	3768	3774	# 10	3990	3996	4002	4008	4014
# 3	4410	4417	4424	4431	4438	# 11	4690	4697	4704	4711	4718
# 4	4445	4452	4459	4466	4473	# 12	4725	4732	4739	4746	4753
# 5	5120	5128	5136	5144	5152	# 13	5440	5448	5456	5464	5472
# 6	5160	5168	5176	5184	5192	# 14	5480	5488	5496	5504	5512
# 7	5850	5859	5868	5877	5886	# 15	6210	6219	6228	6237	6246
# 8	5895	5904	5913	5922	5931	# 16	6255	6264	6273	6282	6291

Summary 5 Answers Chapter 2

Lesson 7

	a	b	c	d	e		a	b	c	d	e
# 1	4320	4326	4332	4338	4344	# 9	4560	4566	4572	4578	4584
# 2	4350	4356	4362	4368	4374	# 10	4590	4596	4602	4608	4614
# 3	5110	5117	5124	5131	5138	# 11	5390	5397	5404	5411	5418
# 4	5145	5152	5159	5166	5173	# 12	5425	5432	5439	5446	5453
# 5	5920	5928	5936	5944	5952	# 13	6240	6248	6256	6264	6272
# 6	5960	5968	5976	5984	5992	# 14	6280	6288	6296	6304	6312
# 7	6750	6759	6768	6777	6786	# 15	7110	7119	7128	7137	7146
# 8	6795	6804	6813	6822	6831	# 16	7155	7164	7173	7182	7191

Lesson 8

	a	b	c	d	e		a	b	c	d	e
# 1	4920	4926	4932	4938	4944	# 9	5160	5166	5172	5178	5184
# 2	4950	4956	4962	4968	4974	# 10	5190	5196	5202	5208	5214
# 3	5810	5817	5824	5831	5838	# 11	6090	6097	6104	6111	6118
# 4	5845	5852	5859	5866	5873	# 12	6125	6132	6139	6146	6153
# 5	6720	6728	6736	6744	6752	# 13	7040	7048	7056	7064	7072
# 6	6760	6768	6776	6784	6792	# 14	7080	7088	7096	7104	7112
# 7	7650	7659	7668	7677	7686	# 15	8010	8019	8028	8037	8046
# 8	7695	7704	7713	7722	7731	# 16	8055	8064	8073	8082	8091

Lesson 9

	a	b	c	d	e		a	b	c	d	e
# 1	5520	5526	5532	5538	5544	# 9	5760	5766	5772	5778	5784
# 2	5550	5556	5562	5568	5574	# 10	5790	5796	5802	5808	5814
# 3	6510	6517	6524	6531	6538	# 11	6790	6797	6804	6811	6818
# 4	6545	6552	6559	6566	6573	# 12	6825	6832	6839	6846	6853
# 5	7520	7528	7536	7544	7552	# 13	7840	7848	7856	7864	7872
# 6	7560	7568	7576	7584	7592	# 14	7880	7888	7896	7904	7912
# 7	8550	8559	8568	8577	8586	# 15	8910	8919	8928	8937	8946
# 8	8595	8604	8613	8622	8631	# 16	8955	8964	8973	8982	8991

Summary 5 Answers Chapter 2

Lesson 10

Test 2 A

#1	2670	#2	2835	#3	3936	#4	4068	#5	5215
#6	5523	#7	6848	#8	6936	#9	8703	#10	8802
#11	2715	#12	2880	#13	4050	#14	4122	#15	5278
#16	5586	#17	6920	#18	7008	#19	8784	#20	8865

Test 2 B

#1	2730	#2	3290	#3	3870	#4	4134	#5	5292
#6	5376	#7	6936	#8	7032	#9	8802	#10	8883
#11	2820	#12	3425	#13	3924	#14	4188	#15	5348
#16	5502	#17	7008	#18	7176	#19	8703	#20	8784

Summary 5

Chapter 3

Division

Numbers 2

Summary 5 Chapter 3 Contents
Division Numbers
Lesson 1 – 10
Basic Problems

Work with lesson 1, read rules and examples, 2) Solve the problems from contents, 3) Check answers in the end of chapter 1, 4) Work with the next lessons 2, 3, 4, 5, …, 5) Solve Test 1 A, 6) Grading Scale: A =10 points, B=9points, C=8 points, D=7 points, F=0-6 points, 7) If you solve 10 problems correctly, you get an A, 8) If have an A, you can move on to next chapter, 9) If your grade is lower than an A, work with the mistakes, then solve Test 1B, 10) When you finish all of the chapters, solve Final Exam in last chapter.

Lesson 1 Division Numbers 1
Solve the problems # 5 (a, b, c, d), # 10 (a, b, c, d)
Lesson 2 Division Numbers 2
Solve the problems # 5 (a, b, c, d), # 10 (a, b, c, d)
Lesson 3 Division Numbers 3
Solve the problems # 5 (a, b, c, d), # 10 (a, b, c, d)
Lesson 4 Division Numbers 4
Solve the problems # 5 (a, b, c, d), # 10(a, b, c, d)
Lesson 5 Division Numbers 5
Solve the problems # 5 (a, b, c, d), # 10 (a, b, c, d)
Lesson 6 Division Numbers 6
Solve the problems # 5 (a, b, c, d), # 10 (a, b, c, d)
Lesson 7 Division Numbers 7
Solve the problems # 5 (a, b, c, d), # 10 (a, b, c, d)
Lesson 8 Division Numbers 8
Solve the problems # 5 (a, b, c, d), # 10 (a, b, c, d)
Lesson 9 Division Numbers 9
Solve the problems # 10 (a, b, c, d), # 20(a, b, c, d)
Lesson 10 Test 3 A, Test 3 B for 50 minutes (10 problems)
Solve Test 3 A # 1, 3, 5, 7, 9, 11, 13, 15, 17, 19
Solve Test 3 B # 1, 3, 5, 7, 9, 11, 13, 15, 17, 19
Answers Chapter 3

Summary 5 Chapter 3 Lesson 1
Division Numbers 1
Every time bring down just one digit
Example 1

105	1) 7 × 1 = 7	Put 1 in quotient	check
7 ⟌ 735	2) 7 – 7 = 0		105
-7	3) Bring down the 3		×7
3	4) 7 × 0 = 0	Put 0 in quotient	35
-0	5) 3 – 0 = 3		000
35	6) Bring down the 5		700
-35	7) 7×5 = 35	Put 5 in quotient	735
0	8) 35 – 35 = 0		

Divide numbers in column

	a	b	c	d
# 1	500 ÷ 5	505 ÷ 5	510 ÷ 5	515 ÷ 5
# 2	520 ÷ 5	525 ÷ 5	530 ÷ 5	535 ÷ 5
# 3	660 ÷ 6	666 ÷ 6	672 ÷ 6	678 ÷ 6
# 4	684 ÷ 6	690 ÷ 6	696 ÷ 6	702 ÷ 6
# 5	840 ÷ 7	847 ÷ 7	854 ÷ 7	861 ÷ 7

Example 2

165	1) 6 × 1 = 6	Put 1 in quotient	check
6 ⟌ 990	2) 9 – 6 = 3		165
-6	3) Bring down the 9		×6
39	4) 6 × 6 = 36	Put 6 in quotient	30
-36	5) 39 – 36 = 3		360
30	6) Bring down the 0		+600
-30	7) 6×5 = 30	Put 5 in quotient	990
0	8) 30 – 30 = 0		

Divide numbers in column

	a	b	c	d
# 6	750 ÷ 5	755 ÷ 5	760 ÷ 5	765 ÷ 5
# 7	770 ÷ 5	775 ÷ 5	780 ÷ 5	785 ÷ 5
# 8	960 ÷ 6	966 ÷ 6	972 ÷ 6	978 ÷ 6
# 9	984 ÷ 6	990 ÷ 6	996 ÷ 6	1002 ÷ 6
# 10	1190 ÷ 7	1197 ÷ 7	1204 ÷ 7	1211 ÷ 7

Summary 5 Chapter 3 Lesson 2

Division Numbers 2

Every time bring down just one digit

Example 1

206	1) $5 \times \underline{2} = 10$	or	1030 | 8
5 | 1030	2) $10 - 10 = 0$		-10 206
- 10	3) Bring down the 3		3
3	4) $5 \times \underline{0} = 0$		-0
-0	5) $3 - 0 = 3$		30
30	6) Bring down the 0		-30
-30	7) $5 \times \underline{6} = 30$		0
0	8) $30 - 30 = 0$		

Divide numbers in column

	a	b	c	d
# 1	1000÷5	1005÷5	1010÷5	1015÷5
# 2	1020÷5	1025÷5	1030÷5	1035÷5
# 3	1260÷6	1266÷6	1272÷6	1278÷6
# 4	1284÷6	1290÷6	1296÷6	1302÷6
# 5	1540÷7	1547÷7	1554÷7	1561÷7

Example 2

285	1) $8 \times \underline{2} = 16$	or	2280 | 8
8 | 2280	2) $22 - 16 = 6$		-16 285
-16	3) Bring down the 8		68
68	4) $8 \times \underline{8} = 64$		-64
- 64	5) $68 - 64 = 4$		40
40	6) Bring down the 0		-40
-40	7) $8 \times \underline{5} = 40$		0
0	8) $40 - 40 = 0$		

Divide numbers in column

	a	b	c	d
# 6	1250÷5	1255÷5	1260÷5	1265÷5
# 7	1270÷5	1275÷5	1280÷5	1285÷5
# 8	1560÷6	1566÷6	1572÷6	1578 ÷6
# 9	1584÷6	1590÷6	1596÷6	1602 ÷6
# 10	1890÷7	1897 ÷7	1904÷7	1911÷7

Summary 5 Chapter 3 Lesson 3
Division Numbers 3
Every time bring down just one digit
Example 1

305	1) $5 \times \underline{3} = 15$	Put 3 in quotient
5⟌1525	2) $15 - 15 = 0$	
-15	3) Bring down the 2	
2	4) $5 \times \underline{0} = 0$	Put 0 in quotient
-0	5) $2 - 0 = 2$	
25	6) Bring down the 5	
-25	7) $5 \times \underline{5} = 25$	Put 5 in quotient
0	8) $25 - 25 = 0$	

Divide numbers in column

	a	b	c	d
# 1	1500÷5	1505÷5	1510÷5	1515÷5
# 2	1520÷5	1525÷5	1530÷5	1535÷5
# 3	1860÷6	1866÷6	1872÷6	1878÷6
# 4	1884÷6	1890÷6	1896÷6	1902÷6
# 5	2240÷7	2247÷7	2254÷7	2261÷7

Example 2

385	1) $8 \times \underline{3} = 24$	Put 3 in quotient
8⟌3080	2) $30 - 24 = 6$	
-24	3) Bring down the 8	
68	4) $8 \times \underline{8} = 64$	Put 8 in quotient
- 64	5) $68 - 64 = 4$	
40	6) Bring down the 0	
-40	7) $8 \times \underline{5} = 40$	Put 5 in quotient
0	8) $40 - 40 = 0$	

Divide numbers in column

	a	b	c	d
# 6	1750÷5	1755÷5	1760 ÷5	1765 ÷5
# 7	1770÷5	1775÷5	1780÷5	1785÷5
# 8	2160÷6	2166÷6	2172÷6	2178÷6
# 9	2184 ÷6	2190÷6	2196÷6	2202÷6
# 10	2590÷7	2597÷7	2604÷7	2611÷7

Summary 5 Chapter 3 Lesson 4
Division Numbers 4
Every time bring down just one digit
Example 1

$\underline{407}$	1) $5 \times \underline{4} = 20$	Put 4 in quotient
$5\overline{)2035}$	2) $20 - 20 = 0$	
$\underline{-20}$	3) Bring down the 3	
3	4) $5 \times \underline{0} = 0$	Put 0 in quotient
$\underline{-0}$	5) $3 - 0 = 3$	
35	6) Bring down the 5	
$\underline{-35}$	7) $5 \times \underline{7} = 35$	Put 7 in quotient
0	8) $35 - 35 = 0$	

Divide numbers in column

	a	b	c	d
# 1	$2000 \div 5$	$2005 \div 5$	$2010 \div 5$	$2015 \div 5$
# 2	$2020 \div 5$	$2025 \div 5$	$2030 \div 5$	$2035 \div 5$
# 3	$2460 \div 6$	$2466 \div 6$	$2472 \div 6$	$2478 \div 6$
# 4	$2484 \div 6$	$2490 \div 6$	$2496 \div 6$	$2502 \div 6$
# 5	$2940 \div 7$	$2947 \div 7$	$2954 \div 7$	$2961 \div 7$

Example 2

$\underline{485}$	1) $8 \times \underline{4} = 32$	Put 4 in quotient
$8\overline{)3880}$	2) $38 - 32 = 6$	
$\underline{-32}$	3) Bring down the 8	
68	4) $8 \times \underline{8} = 64$	Put 8 in quotient
$\underline{-64}$	5) $68 - 64 = 4$	
40	6) Bring down the 0	
$\underline{-40}$	7) $8 \times \underline{5} = 40$	Put 5 in quotient
0	8) $40 - 40 = 0$	

Divide numbers in column

	a	b	c	d
# 6	$2250 \div 5$	$2255 \div 5$	$2260 \div 5$	$2265 \div 5$
# 7	$2270 \div 5$	$2275 \div 5$	$2280 \div 5$	$2285 \div 5$
# 8	$2760 \div 6$	$2766 \div 6$	$2772 \div 6$	$2778 \div 6$
# 9	$2784 \div 6$	$2790 \div 6$	$2796 \div 6$	$2802 \div 6$
# 10	$3290 \div 7$	$3297 \div 7$	$3304 \div 7$	$3311 \div 7$

Summary 5 Chapter 3 Lesson 5
Division Numbers 5
Every time bring down just one digit
Example 1

```
  508
5|2540
 -25
   4
  -0
  40
 -40
   0
```

1) 5 × 5 = 25 Put 5 in quotient
2) 25 – 25 = 0
3) Bring down the 4
4) 5 × 0 = 0 Put 0 in quotient
5) 4 – 0 = 4
6) Bring down the 0
7) 5×8 = 40 Put 8 in quotient
8) 40 – 40 = 0

Divide numbers in column

	a	b	c	d
# 1	2500 ÷ 5	2505 ÷ 5	2510 ÷ 5	2515 ÷ 5
# 2	2520 ÷ 5	2525 ÷ 5	2530 ÷ 5	2535 ÷ 5
# 3	3060 ÷ 6	3066 ÷ 6	3072 ÷ 6	3078 ÷ 6
# 4	3084 ÷ 6	3090 ÷ 6	3096 ÷ 6	3102 ÷ 6
# 5	3640 ÷ 7	3647 ÷ 7	3654 ÷ 7	3661 ÷ 7

Example 2

```
  585
8|4680
 -40
  68
 -64
  40
 -40
   0
```

1) 8 × 5 = 40 Put 5 in quotient
2) 46 – 40 = 6
3) Bring down the 8
4) 8 × 8 = 64 Put 8 in quotient
5) 68 – 64 = 4
6) Bring down the 0
7) 8×5 = 40 Put 5 in quotient
8) 40 – 40 = 0

Divide numbers in column

	a	b	c	d
# 6	2750 ÷ 5	2755 ÷ 5	2760 ÷ 5	2765 ÷ 5
# 7	2770 ÷ 5	2775 ÷ 5	2780 ÷ 5	2785 ÷ 5
# 8	3360 ÷ 6	3366 ÷ 6	3372 ÷ 6	3378 ÷ 6
# 9	3384 ÷ 6	3390 ÷ 6	3396 ÷ 6	3402 ÷ 6
# 10	3990 ÷ 7	3997 ÷ 7	4004 ÷ 7	4011 ÷ 7

Summary 5 Chapter 3 Lesson 6
Division Numbers 6
Every time bring down just one digit
Example 1

609 (quotient)	1) $5 \times \underline{6} = 30$	Put 6 in quotient
5 | 3045	2) $30 - 30 = 0$	
-30	3) Bring down the 4	
4	4) $5 \times \underline{0} = 0$	Put 0 in quotient
-0	5) $4 - 0 = 4$	
45	6) Bring down the 5	
-45	7) $5 \times \underline{9} = 45$	Put 9 in quotient
0	8) $45 - 45 = 0$	

Divide numbers in column

	a	b	c	d
# 1	$3000 \div 5$	$3005 \div 5$	$3010 \div 5$	$3015 \div 5$
# 2	$3020 \div 5$	$3025 \div 5$	$3030 \div 5$	$3035 \div 5$
# 3	$3660 \div 6$	$3666 \div 6$	$3672 \div 6$	$3678 \div 6$
# 4	$3684 \div 6$	$3690 \div 6$	$3696 \div 6$	$3702 \div 6$
# 5	$4340 \div 7$	$4347 \div 7$	$4354 \div 7$	$4361 \div 7$

Example 2

685 (quotient)	1) $8 \times \underline{6} = 48$	Put 6 in quotient
8 | 5480	2) $54 - 48 = 6$	
-48	3) Bring down the 8	
68	4) $8 \times \underline{8} = 64$	Put 8 in quotient
- 64	5) $68 - 64 = 4$	
40	6) Bring down the 0	
-40	7) $8 \times \underline{5} = 40$	Put 5 in quotient
0	8) $40 - 40 = 0$	

Divide numbers in column

	a	b	c	d
# 6	$3250 \div 5$	$3255 \div 5$	$3260 \div 5$	$3265 \div 5$
# 7	$3270 \div 5$	$3275 \div 5$	$3280 \div 5$	$3285 \div 5$
# 8	$3960 \div 6$	$3966 \div 6$	$3972 \div 6$	$3978 \div 6$
# 9	$3984 \div 6$	$3990 \div 6$	$3996 \div 6$	$4002 \div 6$
# 10	$4690 \div 7$	$4697 \div 7$	$4704 \div 7$	$4711 \div 7$

Summary 5 Chapter 3 Lesson 7
Division Numbers 7
Every time bring down just one digit
Example 1

```
   708
5|3540
 -35
    4
   -0
   40
  -40
    0
```

1) $5 \times \underline{7} = 35$ — Put 7 in quotient
2) $35 - 35 = 0$
3) Bring down the 4
4) $5 \times \underline{0} = 0$ — Put 0 in quotient
5) $4 - 0 = 4$
6) Bring down the 0
7) $5 \times \underline{8} = 40$ — Put 8 in quotient
8) $40 - 40 = 0$

Divide numbers in column

	a	b	c	d
# 1	3500 ÷ 5	3505 ÷ 5	3510 ÷ 5	3515 ÷ 5
# 2	3520 ÷ 5	3525 ÷ 5	3530 ÷ 5	3535 ÷ 5
# 3	4260 ÷ 6	4266 ÷ 6	4272 ÷ 6	4278 ÷ 6
# 4	4284 ÷ 6	4290 ÷ 6	4296 ÷ 6	4302 ÷ 6
# 5	5040 ÷ 7	5047 ÷ 7	5054 ÷ 7	5061 ÷ 7

Example 2

```
   785
8|6280
 -56
   68
 - 64
    40
   -40
     0
```

1) $8 \times \underline{7} = 56$ — Put 7 in quotient
2) $62 - 56 = 6$
3) Bring down the 8
4) $8 \times \underline{8} = 64$ — Put 8 in quotient
5) $68 - 64 = 4$
6) Bring down the 0
7) $8 \times \underline{5} = 40$ — Put 5 in quotient
8) $40 - 40 = 0$

Divide numbers in column

	a	b	c	d
# 6	3750 ÷ 5	3755 ÷ 5	3760 ÷ 5	3765 ÷ 5
# 7	3770 ÷ 5	3775 ÷ 5	3780 ÷ 5	3785 ÷ 5
# 8	4560 ÷ 6	4566 ÷ 6	4572 ÷ 6	4578 ÷ 6
# 9	4584 ÷ 6	4590 ÷ 6	4596 ÷ 6	4602 ÷ 6
# 10	5390 ÷ 7	5397 ÷ 7	5404 ÷ 7	5411 ÷ 7

Summary 5 Chapter 3 Lesson 8
Division Numbers 8
Every time bring down just one digit
Example 1

```
   809
5)4045
 -40
    4
   -0
   45
  -45
    0
```

1) $5 \times \underline{8} = 40$ Put 8 in quotient
2) $40 - 40 = 0$
3) Bring down the 4
4) $5 \times \underline{0} = 0$ Put 0 in quotient
5) $4 - 0 = 4$
6) Bring down the 5
7) $5 \times \underline{9} = 45$ Put 9 in quotient
8) $45 - 45 = 0$

Divide numbers in column

	a	b	c	d
# 1	4000 ÷ 5	4005 ÷ 5	4010 ÷ 5	4015 ÷ 5
# 2	4020 ÷ 5	4025 ÷ 5	4030 ÷ 5	4035 ÷ 5
# 3	4860 ÷ 6	4866 ÷ 6	4872 ÷ 6	4878 ÷ 6
# 4	4884 ÷ 6	4890 ÷ 6	4896 ÷ 6	4902 ÷ 6
# 5	5740 ÷ 7	5747 ÷ 7	5754 ÷ 7	5761 ÷ 7

Example 2

```
   875
7)6125
 -56
   52
  - 49
    35
   -35
     0
```

1) $7 \times \underline{8} = 56$ Put 8 in quotient
2) $61 - 56 = 5$
3) Bring down the 2
4) $7 \times \underline{7} = 49$ Put 7 in quotient
5) $52 - 49 = 3$
6) Bring down the 5
7) $7 \times \underline{5} = 35$ Put 5 in quotient
8) $35 - 35 = 0$

Divide numbers in column

	a	b	c	d
# 6	4250 ÷ 5	4255 ÷ 5	4260 ÷ 5	4265 ÷ 5
# 7	4270 ÷ 5	4275 ÷ 5	4280 ÷ 5	4285 ÷ 5
# 8	5160 ÷ 6	5166 ÷ 6	5172 ÷ 6	5178 ÷ 6
# 9	5184 ÷ 6	5190 ÷ 6	5196 ÷ 6	5202 ÷ 6
# 10	6090 ÷ 7	6097 ÷ 7	6104 ÷ 7	6111 ÷ 7

Summary 5 Chapter 3 Lesson 9
Review for Test 3 A
Division Numbers 9
Every time bring down just one digit
Example 1

$\begin{array}{r} 907 \\ 5\overline{)4535} \\ \underline{-45} \\ 3 \\ \underline{-0} \\ 35 \\ \underline{-35} \\ 0 \end{array}$	1) $5 \times \underline{9} = 45$	Put 9 in quotient
	2) $45 - 45 = 0$	
	3) Bring down the 3	
	4) $5 \times \underline{0} = 0$	Put 0 in quotient
	5) $3 - 0 = 3$	
	6) Bring down the 5	
	7) $5 \times \underline{7} = 35$	Put 7 in quotient
	8) $35 - 35 = 0$	

Divide numbers in column

	a	b	c	d
# 1	$4500 \div 5$	$4505 \div 5$	$4510 \div 5$	$4515 \div 5$
# 2	$4520 \div 5$	$4525 \div 5$	$4530 \div 5$	$4535 \div 5$
# 3	$5460 \div 6$	$5466 \div 6$	$5472 \div 6$	$5478 \div 6$
# 4	$5484 \div 6$	$5490 \div 6$	$5496 \div 6$	$5502 \div 6$
# 5	$6440 \div 7$	$6447 \div 7$	$6454 \div 7$	$6461 \div 7$
# 6	$6468 \div 7$	$6475 \div 7$	$6482 \div 7$	$6489 \div 7$
# 7	$7440 \div 8$	$7448 \div 8$	$7456 \div 8$	$7464 \div 8$
# 8	$7472 \div 8$	$7480 \div 8$	$7488 \div 8$	$7496 \div 8$
# 9	$8460 \div 9$	$8469 \div 9$	$8478 \div 9$	$8487 \div 9$
# 10	$8496 \div 9$	$8505 \div 9$	$8514 \div 9$	$8523 \div 9$

Summary 5 Chapter 3 Lesson 9
Review for Test 3 B
Division Numbers 9
Every time bring down just one digit
Example 2

$8\overline{)7880}$ quotient 985	1) $8 \times \underline{9} = 72$	Put 9 in quotient
-72	2) $78 - 72 = 6$	
68	3) Bring down the 8	
- 64	4) $8 \times \underline{8} = 64$	Put 8 in quotient
40	5) $68 - 64 = 4$	
-40	6) Bring down the 0	
0	7) $8\times\underline{5} = 40$	Put 5 in quotient
	8) $40 - 40 = 0$	

Divide numbers in column

	a	b	c	d
# 11	4750 ÷ 5	4755 ÷ 5	4760 ÷ 5	4765 ÷ 5
# 12	4770 ÷ 5	4775 ÷ 5	4780 ÷ 5	4785 ÷ 5
# 13	5760 ÷ 6	5766 ÷ 6	5772 ÷ 6	5778 ÷ 6
# 14	5784 ÷ 6	5790 ÷ 6	5796 ÷ 6	5802 ÷ 6
# 15	6790 ÷ 7	6797 ÷ 7	6804 ÷ 7	6811 ÷ 7
# 16	6818 ÷ 7	6825 ÷ 7	6832 ÷ 7	6839 ÷ 7
# 17	7840 ÷ 8	7848 ÷ 8	7856 ÷ 8	7864 ÷ 8
# 18	7872 ÷ 8	7880 ÷ 8	7888 ÷ 8	7896 ÷ 8
# 19	8910 ÷ 9	8919 ÷ 9	8928 ÷ 9	8937 ÷ 9
# 20	8946 ÷ 9	8955 ÷ 9	8964 ÷ 9	8973 ÷ 9

Summary 5 Chapter 3 Lesson 10

Division Numbers

Test 3 A

Divide numbers

# 1	2730 ÷ 5	# 2	2890 ÷ 5
# 3	3870 ÷ 6	# 4	4134 ÷ 6
# 5	5292 ÷ 7	# 6	5383 ÷ 7
# 7	6936 ÷ 8	# 8	7024 ÷ 8
# 9	8802 ÷ 9	# 10	8883 ÷ 9
# 11	2670 ÷ 5	# 12	2835 ÷ 5
# 13	3936 ÷ 6	# 14	4068 ÷ 6
# 15	5215 ÷ 7	# 16	5523 ÷ 7
# 17	6672 ÷ 8	# 18	6848 ÷ 8
# 19	8613 ÷ 9	# 20	8874 ÷ 9

Summary 5 Chapter 3 Lesson 10

Division Numbers

Test 3 B

Divide numbers

# 1	**$2820 \div 5$**	**# 2**	**$2935 \div 5$**
# 3	**$3924 \div 6$**	**# 4**	**$4188 \div 6$**
# 5	**$5355 \div 7$**	**# 6**	**$5572 \div 7$**
# 7	**$7000 \div 8$**	**# 8**	**$7168 \div 8$**
# 9	**$8703 \div 9$**	**# 10**	**$8802 \div 9$**
# 11	**$2715 \div 5$**	**# 12**	**$2880 \div 5$**
# 13	**$4050 \div 6$**	**# 14**	**$4122 \div 6$**
# 15	**$5278 \div 7$**	**# 16**	**$5586 \div 7$**
# 17	**$6768 \div 8$**	**# 18**	**$6920 \div 8$**
# 19	**$8784 \div 9$**	**# 20**	**$8901 \div 9$**

Summary 5 Answers Chapter 3

Lesson 1a	b	c	d		a	b	c	d
# 1 100	101	102	103	# 6	150	151	152	153
# 2 104	105	106	107	# 7	154	155	156	157
# 3 110	111	112	113	# 8	160	161	162	163
# 4 114	115	116	117	# 9	164	165	166	167
# 5 120	121	122	123	#10	170	171	172	173
Lesson 2a	**b**	**c**	**d**		**a**	**b**	**c**	**d**
# 1 200	201	202	203	# 6	250	251	252	253
# 2 204	205	206	207	# 7	254	255	256	257
# 3 210	211	212	213	# 8	260	261	262	263
# 4 214	215	216	217	# 9	264	265	266	267
# 5 220	221	222	223	#10	270	271	272	273
Lesson 3a	**b**	**c**	**d**		**a**	**b**	**c**	**d**
# 1 300	301	302	303	# 6	350	351	352	353
# 2 304	305	306	307	# 7	354	355	356	357
# 3 310	311	312	313	# 8	360	361	362	363
# 4 314	315	316	317	# 9	364	365	366	367
# 5 320	321	322	323	#10	370	371	372	373
Lesson 4a	**b**	**c**	**d**		**a**	**b**	**c**	**d**
# 1 400	401	402	403	# 6	450	451	452	453
# 2 404	405	406	407	# 7	454	455	456	457
# 3 410	411	412	413	# 8	460	461	462	463
# 4 414	415	416	417	# 9	464	465	466	467
# 5 420	421	422	423	#10	470	471	472	473
Lesson 5a	**b**	**c**	**d**		**a**	**b**	**c**	**d**
# 1 500	501	502	503	# 6	550	551	552	553
# 2 504	505	506	507	# 7	554	555	556	557
# 3 510	511	512	513	# 8	560	561	562	563
# 4 514	515	516	517	# 9	564	565	566	567
# 5 520	521	522	523	#10	570	571	572	573
Lesson 6a	**b**	**c**	**d**		**a**	**b**	**c**	**d**
# 1 600	601	602	603	# 6	650	651	652	653
# 2 604	605	606	607	# 7	654	655	656	657
# 3 610	611	612	613	# 8	660	661	662	663
# 4 614	615	616	617	# 9	664	665	666	667
# 5 620	621	622	623	#10	670	671	672	673

Summary 5 Answers Chapter 3

Lesson 7

	a	b	c	d		a	b	c	d
# 1	700	701	702	703	# 6	750	751	752	753
# 2	704	705	706	707	# 7	754	755	756	757
# 3	710	711	712	713	# 8	760	761	762	763
# 4	714	715	716	717	# 9	764	765	766	767
# 5	720	721	722	723	#10	770	771	772	773

Lesson 8

	a	b	c	d		a	b	c	d
# 1	800	801	802	803	# 6	850	851	852	853
# 2	804	805	806	807	# 7	854	855	856	857
# 3	810	811	812	813	# 8	860	861	862	863
# 4	814	815	816	817	# 9	864	865	866	867
# 5	820	821	822	823	#10	870	871	872	873

Lesson 9

	a	b	c	d		a	b	c	d
# 1	900	901	902	903	#11	950	951	952	953
# 2	904	905	906	907	#12	954	955	956	957
# 3	910	911	912	913	#13	960	961	962	963
# 4	914	915	916	917	#14	964	965	966	967
# 5	920	921	922	923	#15	970	971	972	973
# 6	924	925	926	927	#16	974	975	976	977
# 7	930	931	932	933	#17	980	981	982	983
# 8	934	935	936	937	#18	984	985	986	987
# 9	940	941	942	943	#19	990	991	992	993
# 10	944	945	946	947	#20	994	995	996	997

Lesson 10

Test 3 A

#1	546	#2	578	#3	645	#4	689	#5	756
#6	769	#7	867	#8	878	#9	978	#10	987
#11	534	#12	567	#13	656	#14	678	#15	745
#16	789	#17	834	#18	856	#19	957	#20	986

Test 3 B

#1	564	#2	587	#3	654	#4	698	#5	765
#6	796	#7	875	#8	896	#9	967	#10	978
#11	543	#12	576	#13	675	#14	687	#15	754
#16	798	#17	846	#18	865	#19	976	#20	989

Summary 5

Chapter 4 Four Operations with numbers 2

Summary 5 Chapter 4 Contents
Four operations with four-digit numbers
Lesson 1 - 12
Basic Problems

Work with lesson 1, read rules and examples, 2) Solve the problems from contents, 3) Check answers in the end of chapter 1, 4) Work with the next lessons 2, 3, 4, 5, …, 5) Solve Test 1 A, 6) Grading Scale: A =10 points, B=9points, C=8 points, D=7 points, F=0-6 points, 7) If you solve 10 problems correctly, you get an A, 8) If have an A, you can move on to next chapter, 9) If your grade is lower than an A, work with the mistakes, then solve Test 1B, 10) When you finish all of the chapters, solve Final Exam in last chapter.

Summary 5 Chapter 4 Lesson 1

Multiplication numbers

Examples

To multiply numbers, work in any order

Example a $10 \times 8 \times 9 = 80 \times 9 = 720$

Example b $10 \times 8 \times 9 = 10 \times 72 = 720$

Review 1 A

Multiply numbers

	a	b	c
# 1	$8 \times 2 \times 1$	$8 \times 4 \times 1$	$8 \times 6 \times 1$
# 2	$8 \times 2 \times 2$	$8 \times 4 \times 2$	$8 \times 6 \times 2$
# 3	$8 \times 2 \times 3$	$8 \times 4 \times 3$	$8 \times 6 \times 3$
# 4	$8 \times 2 \times 4$	$8 \times 4 \times 4$	$8 \times 6 \times 4$
# 5	$8 \times 2 \times 5$	$8 \times 4 \times 5$	$8 \times 6 \times 5$
# 6	$8 \times 2 \times 6$	$8 \times 4 \times 6$	$8 \times 6 \times 6$
# 7	$8 \times 2 \times 7$	$8 \times 4 \times 7$	$8 \times 6 \times 7$
# 8	$8 \times 2 \times 8$	$8 \times 4 \times 8$	$8 \times 6 \times 8$
# 9	$8 \times 2 \times 9$	$8 \times 4 \times 9$	$8 \times 6 \times 9$
# 10	$8 \times 2 \times 10$	$8 \times 4 \times 10$	$8 \times 6 \times 10$
# 11	$9 \times 3 \times 1$	$9 \times 5 \times 1$	$9 \times 7 \times 1$
# 12	$9 \times 3 \times 2$	$9 \times 5 \times 2$	$9 \times 7 \times 2$
# 13	$9 \times 3 \times 3$	$9 \times 5 \times 3$	$9 \times 7 \times 3$
# 14	$9 \times 3 \times 4$	$9 \times 5 \times 4$	$9 \times 7 \times 4$
# 15	$9 \times 3 \times 5$	$9 \times 5 \times 5$	$9 \times 7 \times 5$
# 16	$9 \times 3 \times 6$	$9 \times 5 \times 6$	$9 \times 7 \times 6$
# 17	$9 \times 3 \times 7$	$9 \times 5 \times 7$	$9 \times 7 \times 7$
# 18	$9 \times 3 \times 8$	$9 \times 5 \times 8$	$9 \times 7 \times 8$
# 19	$9 \times 3 \times 9$	$9 \times 5 \times 9$	$9 \times 7 \times 9$
# 20	$9 \times 3 \times 10$	$9 \times 5 \times 10$	$9 \times 7 \times 10$

Summary 5 Chapter 4 Lesson 1
Multiplication and Division numbers
To multiply and divide numbers work from left to right

Example a $10 \times 81 \div 9 = 810 \div 9 = 90$

Example b $10 \times 72 \div 9 = 720 \div 9 = 80$

Review 1 B

Multiply and Divide numbers

	a	b	c
# 1	$8 \times 2 \div 2$	$8 \times 4 \div 4$	$8 \times 6 \div 6$
# 2	$8 \times 4 \div 2$	$8 \times 8 \div 4$	$8 \times 12 \div 6$
# 3	$8 \times 6 \div 2$	$8 \times 12 \div 4$	$8 \times 18 \div 6$
# 4	$8 \times 8 \div 2$	$8 \times 16 \div 4$	$8 \times 24 \div 6$
# 5	$8 \times 10 \div 2$	$8 \times 20 \div 4$	$8 \times 30 \div 6$
# 6	$8 \times 12 \div 2$	$8 \times 24 \div 4$	$8 \times 36 \div 6$
# 7	$8 \times 14 \div 2$	$8 \times 28 \div 4$	$8 \times 42 \div 6$
# 8	$8 \times 16 \div 2$	$8 \times 32 \div 4$	$8 \times 48 \div 6$
# 9	$8 \times 18 \div 2$	$8 \times 36 \div 4$	$8 \times 54 \div 6$
# 10	$8 \times 20 \div 2$	$8 \times 40 \div 4$	$8 \times 60 \div 6$
# 11	$9 \times 3 \div 3$	$9 \times 5 \div 5$	$9 \times 7 \div 7$
# 12	$9 \times 6 \div 3$	$9 \times 10 \div 5$	$9 \times 14 \div 7$
# 13	$9 \times 9 \div 3$	$9 \times 15 \div 5$	$9 \times 21 \div 7$
# 14	$9 \times 12 \div 3$	$9 \times 20 \div 5$	$9 \times 28 \div 7$
# 15	$9 \times 15 \div 3$	$9 \times 25 \div 5$	$9 \times 35 \div 7$
# 16	$9 \times 18 \div 3$	$9 \times 30 \div 5$	$9 \times 42 \div 7$
# 17	$9 \times 21 \div 3$	$9 \times 35 \div 5$	$9 \times 49 \div 7$
# 18	$9 \times 24 \div 3$	$9 \times 40 \div 5$	$9 \times 56 \div 7$
# 19	$9 \times 27 \div 3$	$9 \times 45 \div 5$	$9 \times 63 \div 7$
# 20	$9 \times 30 \div 3$	$9 \times 50 \div 5$	$9 \times 70 \div 7$

Summary 5 Chapter 4 Lesson 2

Division and Multiplication numbers

To divide and multiply numbers work from left to right

Example a $80 \div 8 \times 10 = 10 \times 10 = 100$

Example b $90 \div 9 \times 10 = 10 \times 10 = 100$

Review 1A

Divide and multiply numbers

	a	b	c
# 1	$2 \div 2 \times 8$	$4 \div 4 \times 8$	$6 \div 6 \times 8$
# 2	$4 \div 2 \times 8$	$8 \div 4 \times 8$	$12 \div 6 \times 8$
# 3	$6 \div 2 \times 8$	$12 \div 4 \times 8$	$18 \div 6 \times 8$
# 4	$8 \div 2 \times 8$	$16 \div 4 \times 8$	$24 \div 6 \times 8$
# 5	$10 \div 2 \times 8$	$20 \div 4 \times 8$	$30 \div 6 \times 8$
# 6	$12 \div 2 \times 8$	$24 \div 4 \times 8$	$36 \div 6 \times 8$
# 7	$14 \div 2 \times 8$	$28 \div 4 \times 8$	$42 \div 6 \times 8$
# 8	$16 \div 2 \times 8$	$32 \div 4 \times 8$	$48 \div 6 \times 8$
# 9	$18 \div 2 \times 8$	$36 \div 4 \times 8$	$54 \div 6 \times 8$
# 10	$20 \div 2 \times 8$	$40 \div 4 \times 8$	$60 \div 6 \times 8$
# 11	$3 \div 3 \times 9$	$5 \div 5 \times 9$	$7 \div 7 \times 9$
# 12	$6 \div 3 \times 9$	$10 \div 5 \times 9$	$14 \div 7 \times 9$
# 13	$9 \div 3 \times 9$	$15 \div 5 \times 9$	$21 \div 7 \times 9$
# 14	$12 \div 3 \times 9$	$20 \div 5 \times 9$	$28 \div 7 \times 9$
# 15	$15 \div 3 \times 9$	$25 \div 5 \times 9$	$35 \div 7 \times 9$
# 16	$18 \div 3 \times 9$	$30 \div 5 \times 9$	$42 \div 7 \times 9$
# 17	$21 \div 3 \times 9$	$35 \div 5 \times 9$	$49 \div 7 \times 9$
# 18	$24 \div 3 \times 9$	$40 \div 5 \times 9$	$56 \div 7 \times 9$
# 19	$27 \div 3 \times 9$	$45 \div 5 \times 9$	$63 \div 7 \times 9$
# 20	$30 \div 3 \times 9$	$50 \div 5 \times 9$	$70 \div 7 \times 9$

Summary 5 Chapter 4 Lesson 2

Division numbers

To divide numbers, work from left to right

Example a 6400 ÷ 80 ÷ 8 = 80 ÷ 8 = 10

Example b 8100 ÷ 90 ÷ 9 = 90 ÷ 9 = 10

Review 1 B

Divide numbers

	a	b	c
# 1	4 ÷ 2 ÷ 2	16 ÷ 4 ÷ 4	36 ÷ 6 ÷ 6
# 2	8 ÷ 2 ÷ 2	32 ÷ 4 ÷ 4	72 ÷ 6 ÷ 6
# 3	12 ÷ 2 ÷ 2	48 ÷ 4 ÷ 4	108 ÷ 6 ÷ 6
# 4	16 ÷ 2 ÷ 2	64 ÷ 4 ÷ 4	144 ÷ 6 ÷ 6
# 5	20 ÷ 2 ÷ 2	80 ÷ 4 ÷ 4	180 ÷ 6 ÷ 6
# 6	24 ÷ 2 ÷ 2	96 ÷ 4 ÷ 4	216 ÷ 6 ÷ 6
# 7	28 ÷ 2 ÷ 2	112 ÷ 4 ÷ 4	252 ÷ 6 ÷ 6
# 8	32 ÷ 2 ÷ 2	128 ÷ 4 ÷ 4	288 ÷ 6 ÷ 6
# 9	36 ÷ 2 ÷ 2	144 ÷ 4 ÷ 4	324 ÷ 6 ÷ 6
# 10	40 ÷ 2 ÷ 2	160 ÷ 4 ÷ 4	360 ÷ 6 ÷ 6
# 11	9 ÷ 3 ÷ 3	25 ÷ 5 ÷ 5	49 ÷ 7 ÷ 7
# 12	18 ÷ 3 ÷ 3	50 ÷ 5 ÷ 5	98 ÷ 7 ÷ 7
# 13	27 ÷ 3 ÷ 3	75 ÷ 5 ÷ 5	147 ÷ 7 ÷ 7
# 14	36 ÷ 3 ÷ 3	100 ÷ 5 ÷ 5	196 ÷ 7 ÷ 7
# 15	45 ÷ 3 ÷ 3	125 ÷ 5 ÷ 5	245 ÷ 7 ÷ 7
# 16	54 ÷ 3 ÷ 3	150 ÷ 5 ÷ 5	294 ÷ 7 ÷ 7
# 17	63 ÷ 3 ÷ 3	175 ÷ 5 ÷ 5	343 ÷ 7 ÷ 7
# 18	72 ÷ 3 ÷ 3	200 ÷ 5 ÷ 5	392 ÷ 7 ÷ 7
# 19	81 ÷ 3 ÷ 3	225 ÷ 5 ÷ 5	441 ÷ 7 ÷ 7
# 20	90 ÷ 3 ÷ 3	250 ÷ 5 ÷ 5	490 ÷ 7 ÷ 7

Summary 5 Chapter 4 Lesson 3

Multiplication and Addition numbers

To multiply and add numbers always first multiply after that add

1 2

a) $900 \times 5 + 500 = 4500 + 500 = 5000$

2 1

b) $500 + 700 \times 5 = 500 + 3500 = 4000$

Review 1 A

Evaluate

	a	b
# 1	$100 \times 5 + 100$	$450 \times 5 + 750$
# 2	$200 \times 6 + 800$	$500 \times 6 + 1000$
# 3	$300 \times 7 + 900$	$550 \times 7 + 150$
# 4	$400 \times 8 + 800$	$600 \times 8 + 200$
# 5	$500 \times 9 + 500$	$650 \times 9 + 150$
# 6	$50 + 150 \times 5$	$500 + 700 \times 5$
# 7	$800 + 200 \times 6$	$500 + 750 \times 6$
# 8	$250 + 250 \times 7$	$400 + 800 \times 7$
# 9	$600 + 300 \times 8$	$200 + 850 \times 8$
# 10	$850 + 350 \times 9$	$900 + 900 \times 9$

Multiplication and Subtraction numbers

To multiply and subtract numbers always first multiply after that subtract

1 2

a) $90 \times 80 - 170 = 7200 - 200 = 700$

2 1

b) $9200 - 90 \times 80 = 9200 - 7200 = 2000$

Review 1 B

Evaluate

	a	b
# 11	$10 \times 90 - 100$	$60 \times 50 - 1000$
# 12	$20 \times 90 - 800$	$70 \times 40 - 800$
# 13	$30 \times 80 - 400$	$80 \times 30 - 400$
# 14	$40 \times 70 - 800$	$90 \times 20 - 800$
# 15	$50 \times 60 - 1000$	$90 \times 10 - 100$
# 16	$9900 - 90 \times 10$	$9000 - 50 \times 60$
# 17	$8800 - 90 \times 20$	$7800 - 40 \times 70$
# 18	$7400 - 80 \times 30$	$6400 - 30 \times 80$
# 19	$6800 - 70 \times 40$	$5800 - 20 \times 90$
# 20	$6000 - 60 \times 50$	$9900 - 10 \times 90$

Summary 5 Chapter 4 Lesson 4
Division and Addition numbers
To divide and add numbers always first divide after that add

1 2
a) $4200 \div 5 + 160 = 840 + 160 = 1000$

2 1
b) $100 + 4500 \div 5 = 100 + 900 = 1000$

Review 1 A
Evaluate

	a	b
# 1	4500 ÷ 5 + 500	2000 ÷ 5 + 600
# 2	4800 ÷ 6 + 400	3600 ÷ 6 + 400
# 3	5600 ÷ 7 + 300	4200 ÷ 7 + 1400
# 4	6400 ÷ 8 + 200	5600 ÷ 8 + 1300
# 5	7200 ÷ 9 + 100	6300 ÷ 9 + 2300
# 6	200 + 4500 ÷ 5	100 + 4000 ÷ 5
# 7	300 + 4200 ÷ 6	600 + 5400 ÷ 6
# 8	1400 + 5600 ÷ 7	2100 + 6300 ÷ 7
# 9	2400 + 6400 ÷ 8	1600 + 7200 ÷ 8
# 10	900 + 7200 ÷ 9	900 + 8100 ÷ 9

Division and Subtraction numbers
To divide and subtract numbers always first divide after that subtract

1 2
a) $2400 \div 80 - 30 = 30 - 30 = 0$

2 1
b) $2020 - 1600 \div 80 = 2020 - 20 = 2000$

Review 1 B
Evaluate

	a	b
# 11	5000 ÷ 10 - 100	1000 ÷ 50 - 10
# 12	6000 ÷ 20 – 30	1800 ÷ 60 - 10
# 13	6000 ÷ 30 - 100	3500 ÷ 70 - 20
# 14	8000 ÷ 40 - 100	4000 ÷ 80 - 20
# 15	5000 ÷ 50 – 60	3600 ÷ 90 - 30
# 16	1000 - 500 ÷ 10	1000 - 500 ÷ 50
# 17	1200 - 600 ÷ 20	1200 - 600 ÷ 60
# 18	1700 - 600 ÷ 30	1400 - 700 ÷ 70
# 19	1800 - 800 ÷ 40	1600 - 800 ÷ 80
# 20	1500 - 1000 ÷ 50	1800 - 900 ÷ 90

Summary 5 Chapter 4 Lesson 5
Summary 1
Four-digit numbers

Four-digit numbers content four digits: 1000, 1001, … 9999
Least four-digit number: 1000
Greatest four-digit number: 9999
Digits count from right to the left: ones, tens, hundreds, thousands
2468 = 2000 + 400 + 60 + 8

Four-digit numbers greater than one, two, three-digit numbers
One-digit numbers always less than two, three, four-digit numbers
To compare numbers, compare thousands, hundreds, tens, and ones:

1234 < 1239 　 1238 > 1230 　 1234=1234

Two numbers can add in any order. The sum will be the same.

2000 + 3000 = 5000 　 3000 + 2000 = 5000

The sum of a number and zero is the same number

5000 + 0 = 5000 　 0 + 5000 = 5000

Difference any number and zero is the same number

5000 – 0 = 5000

The difference two the same number is zero

5000 – 5000 = 0

Adding four digits numbers:
Add the ones. Add the tens. Add the hundreds. Add the thousands.
Subtracting four digits numbers: Subtract the ones. Subtract the tens.
Subtract the hundreds. Subtract the thousands.

		1	8 13 14 16
3246	9678	5678	9 4 5 6
+ 4351	- 5432	+1050	- 3 9 8 7
7597	4246	6728	5 4 6 9

Summary 5 Chapter 4 Lesson 5

Summary 2

Multiplication numbers

Third operation with numbers is multiplication × or ·

Multiplication is the operation of repeated addition

$2\times 3 = 2 + 2 + 2 = 6$ or $2 \cdot 3 = 2 + 2 + 2 = 6$

First number in multiplication called the factors 2

Second number in multiplication called the factor 3

Result of multiplication called the product 6

Properties of multiplication

Two numbers can multiply in any order. Product will the same

$2 \times 3 = 2 + 2 + 2 = 6$ $3 \times 2 = 3 + 3 = 6$

The product of a number and zero is zero

$5 \times 0 = 0$ $0 \times 5 = 0$

The product of a number and one is the same number

$5 \times 1 = 5$ $1 \times 5 = 5$

Division numbers

Fourth operation with numbers is division ÷ or :

$6 \div 2 = 3$ $6 : 3 = 2$

Division means separated number by some equal groups

First number in division is called dividend 6

Second number in division is called divisor 2

Result of division is called quotient 3

To check multiplication use division

$2\times3=6$ Check $6\div2=3$ or $6\div3=2$

To check division use multiplication and division

$6\div2=3$ Check $2\times3=6$ or $6\div3=2$

Properties of division

Any number divided by one is the same number $5 \div 1 = 5$

Any number divided by the itself is one $5 \div 5 = 1$

Zero divided by any number is zero $0 \div 5 = 0$

Division by zero is not allowed $5 \div 0$ No number

Summary 5 Chapter 4 Lesson 5

Summary 3 Multiplication numbers

To multiply numbers, work in any order

$9 \times 8 \times 7 = 72 \times 7 = 504$ $\quad$ $9 \times 8 \times 7 = 9 \times 56 = 504$

Multiplication and Division numbers

To multiply and divide numbers work from left to right

$9 \times 6 \div 3 = 54 \div 3 = 18$ $\quad$ $4 \times 81 \div 9 = 324 \div 9 = 36$

Division and Multiplication numbers

To divide and multiply numbers work from left to right

$24 \div 6 \times 2 = 4 \times 2 = 8$ $\quad$ $81 \div 9 \times 4 = 9 \times 4 = 36$

Division numbers

To divide numbers, work from left to right

$24 \div 4 \div 2 = 6 \div 2 = 3$ $\quad$ $810 \div 9 \div 3 = 90 \div 3 = 30$

Multiplication and Addition numbers

To multiply and add numbers always first multiply than add

$9 \times 6 + 9 = 54 + 9 = 63$ $\quad$ $9 + 9 \times 9 = 9 + 81 = 90$

Multiplication and Subtraction numbers

To multiply and subtract numbers always first multiply than subtract.

$9 \times 9 - 9 = 81 - 9 = 72$ $\quad$ $90 - 9 \times 9 = 90 - 81 = 9$

Division and Addition numbers

To divide and add numbers always first divide than add

$50 \div 5 + 39 = 10 + 39 = 49$ $\quad$ $88 + 171 \div 9 = 88 + 19 = 107$

Division and Subtraction numbers

To divide and subtract numbers always first divide than subtract

$108 \div 6 - 9 = 18 - 9 = 9$ $\quad$ $500 - 126 \div 7 = 500 - 18 = 482$

Do multiplication or division before addition or subtraction.

Summary 4 Finding unknown number

Multiplication numbers: First number in multiplication called the factors 2. Second number in multiplication called the factor 3.

Result of multiplication called the product 6

Factor $\times$ Factor = Product $\quad$ $2 \times 3 = 6$

First Factor = Product $\div$ Second Factor $? \times 3 = 6$ $\quad ? = 6 \div 3 \quad ? = 2$

Second Factor = Product $\div$ First Factor $2 \times ? = 6$ $\quad ? = 6 \div 2 \quad ? = 3$

Division numbers: First number in division is called dividend 6

Second number in division is called divisor 2

Result of division is called quotient 3

Dividend $\div$ Divisor = Quotient $6 \div 2 = 3$

Divisor = Dividend $\div$ Quotient $6 \div ? = 3 \quad ? = 6 \div 3 \quad ? = 2$

Dividend = Quotient $\times$ Divisor $? \div 2 = 3 \quad ? = 2 \times 3 \quad ? = 6$

Summary 5 Chapter 4 Lesson 6
Multiplication Numbers
Review 1 A
Multiply numbers

# 1 105×5	# 2 106×5	# 3 107×5	# 4 108×5
# 5 205×6	# 6 206×6	# 7 207×6	# 8 208×6
# 9 305×7	# 10 306×7	# 11 307×7	# 12 308×7
# 13 405×8	# 14 406×8	# 15 407×8	# 16 408×8
# 17 505×9	# 18 506×9	# 19 507×9	# 20 508×9

Review 1 B
Multiply numbers

# 1 605×5	# 2 606×5	# 3 607×5	# 4 608×5
# 5 705×6	# 6 706×6	# 7 707×6	# 8 708×6
# 9 805×7	# 10 806×7	# 11 807×7	# 12 808×7
# 13 905×8	# 14 906×8	# 15 907×8	# 16 908×8
# 17 509×9	# 18 609×9	# 19 709×9	# 20 809×9

Summary 5 Chapter 4 Lesson 6
Division numbers
Review 2 A
Divide numbers

# 1	525 ÷ 5	# 2	530 ÷ 5
# 3	1230 ÷ 6	# 4	1236 ÷ 6
# 5	2135 ÷ 7	# 6	2142 ÷ 7
# 7	3240 ÷ 8	# 8	3248 ÷ 8
# 9	4545 ÷ 9	# 10	4554 ÷ 9
# 11	3025÷5	# 12	3030 ÷ 5
# 13	4230 ÷ 6	# 14	4236 ÷ 6
# 15	5635 ÷ 7	# 16	5642 ÷ 7
# 17	7240 ÷ 8	# 18	7248 ÷ 8
# 19	4581 ÷ 9	# 20	4572 ÷ 9

Review 2 B
Divide numbers

# 1	535 ÷ 5	# 2	540 ÷ 5
# 3	1242 ÷ 6	# 4	1248÷6
# 5	2149 ÷ 7	# 6	2156 ÷ 7
# 7	3256 ÷ 8	# 8	3264 ÷ 8
# 9	4563 ÷ 9	# 10	4572 ÷ 9
# 11	3035 ÷ 5	# 12	3040 ÷ 5
# 13	4242 ÷ 6	# 14	4248 ÷ 6
# 15	5649 ÷ 7	# 16	5656 ÷ 7
# 17	7256 ÷ 8	# 18	7264 ÷ 8
# 19	4563 ÷ 9	# 20	4554 ÷ 9

Summary 5 Chapter 4 Lesson 7
Multiplication two-digit numbers
Example

$$\begin{array}{r} 39 \\ \times \underline{52} \\ 78 \\ +\underline{1950} \\ 2028 \end{array}$$

1) 52=50+2

2) $\begin{array}{r} 39 \\ \underline{\times 2} \\ 78 \end{array}$

3) $\begin{array}{r} 39 \\ \underline{\times 50} \\ 1950 \end{array}$

Review 1 A
Multiply numbers

1 $\begin{array}{r} 11 \\ \times \underline{10} \end{array}$ #2 $\begin{array}{r} 12 \\ \times \underline{20} \end{array}$ #3 $\begin{array}{r} 13 \\ \times \underline{30} \end{array}$ #4 $\begin{array}{r} 14 \\ \times \underline{40} \end{array}$ # 5 $\begin{array}{r} 15 \\ \times \underline{50} \end{array}$

6 $\begin{array}{r} 21 \\ \times \underline{21} \end{array}$ #7 $\begin{array}{r} 22 \\ \times \underline{22} \end{array}$ # 8 $\begin{array}{r} 23 \\ \times \underline{23} \end{array}$ #9 $\begin{array}{r} 24 \\ \times \underline{24} \end{array}$ #10 $\begin{array}{r} 25 \\ \times \underline{25} \end{array}$

11 $\begin{array}{r} 26 \\ \times \underline{15} \end{array}$ #12 $\begin{array}{r} 27 \\ \times \underline{14} \end{array}$ #13 $\begin{array}{r} 28 \\ \times \underline{13} \end{array}$ # 14 $\begin{array}{r} 29 \\ \times \underline{12} \end{array}$ #15 $\begin{array}{r} 30 \\ \times \underline{11} \end{array}$

#16 $\begin{array}{r} 31 \\ \times \underline{25} \end{array}$ #17 $\begin{array}{r} 32 \\ \times \underline{24} \end{array}$ #18 $\begin{array}{r} 33 \\ \times \underline{23} \end{array}$ #19 $\begin{array}{r} 34 \\ \times \underline{22} \end{array}$ #20 $\begin{array}{r} 35 \\ \times \underline{21} \end{array}$

Summary 5 Chapter 4 Lesson 7
Multiplication two-digit numbers

Example

4	1) $2 \times 9 = 18$
1	2) $2 \times 3 = 6$
39	3) $6 + 1 = 7$
$\times \underline{52}$	4) $5 \times 9 = 45$
78	5) $5 \times 3 = 15$
$+\underline{1950}$	6) $15 + 4 = 19$
2028	7) $78 + 1950 = 2028$

Review 1 B
Multiply numbers

# 1 $41 \times \underline{35}$	#2 $42 \times \underline{34}$	#3 $43 \times \underline{33}$	#4 $44 \times \underline{32}$	#5 $45 \times \underline{31}$
# 6 $51 \times \underline{45}$	#7 $52 \times \underline{44}$	#8 $53 \times \underline{43}$	#9 $54 \times \underline{42}$	#10 $55 \times \underline{41}$
# 11 $61 \times \underline{55}$	#12 $62 \times \underline{54}$	#13 $63 \times \underline{53}$	#14 $64 \times \underline{52}$	#15 $65 \times \underline{51}$
# 16 $71 \times \underline{65}$	#17 $72 \times \underline{64}$	#18 $73 \times \underline{63}$	#19 $74 \times \underline{62}$	#20 $75 \times \underline{61}$

Summary 5 Chapter 4 Lesson 8
Division by two-digit numbers
Every time bring down just one digit
Example 1

37 < 54   54 × 6 = 324   54 × 7 = 378   378 > 372

```
     69     1) 54 × 6 = 324          or   3726 | 54
54 | 3726   2) 372 – 324 = 48             -324   69
   -324     3) Bring down the 6            486
     486    4) 54 × 9 = 486               - 486
    -486    5) 486 – 486 = 0                  0
```

Review 1 A
Divide numbers in column

# 1	110 ÷ 11	#2	390 ÷ 15	#3	1435 ÷ 35	#4	3355 ÷ 55
# 5	240 ÷ 12	#6	378 ÷ 14	#7	1428 ÷ 34	#8	3348 ÷ 54
# 9	390 ÷ 13	#10	364 ÷ 13	#11	1419 ÷ 33	#12	3339 ÷ 53
# 13	560 ÷ 14	#14	348 ÷ 12	#15	1408 ÷ 32	#16	3328 ÷ 52
# 17	750 ÷ 15	#18	330 ÷ 11	#19	1395 ÷ 31	#20	3315 ÷ 51

Example 2

40 < 59   59×6 = 354   59×7 = 413   413 > 407

```
     69     1) 59 × 6 = 354          or   4071 | 59
59 | 4071   2) 407 – 354 = 53             -354   69
   -354     3) 69 × 9 = 531                531
     531    4) 531 – 531 = 0              -531
    -531
```

Review 1 B
Divide numbers in column

# 1	6075 ÷ 75	#2	1550 ÷ 50	#3	3570 ÷ 51	#4	6390 ÷ 71
# 5	6068 ÷ 74	# 6	1568 ÷ 49	#7	3588 ÷ 52	#8	6408 ÷ 72
# 9	6059 ÷ 73	#10	1584 ÷ 48	#11	3604 ÷ 53	#12	6424 ÷ 73
# 13	6048 ÷ 72	#14	1598 ÷ 47	#15	3618 ÷ 54	#16	6438 ÷ 74
# 17	6035 ÷ 71	#18	1610 ÷ 46	#19	3630 ÷ 55	#20	6450 ÷ 75

Summary 5 Chapter 4 Lesson 9
Four operations with four-digit numbers
Review 4 A
Perform operations with numbers

# 1	$98\times 4\times 5$	# 11	$96 \times 8 \times 5$
# 2	$96\times 40 \div 8$	# 12	$98\times 45 \div 9$
# 3	$35 \div 7\times 986$	# 13	$30 \div 6\times 978$
# 4	$7848 \div 9 \div 8$	# 14	$6048 \div 8 \div 7$
# 5	$985\times 8 + 120$	# 15	$120 + 976\times 5$
# 6	$95\times 98 - 10$	# 16	$96\times 95 - 20$
# 7	$8200 - 85\times 96$	# 17	$8000 - 84\times 95$
# 8	$7264\div 8 + 92$	# 18	$5436\div 6 + 94$
# 9	$8544\div 96 - 19$	# 19	$8526\div 98 - 27$
# 10	$9598 - 9310\div 95$	# 20	$9396 - 9120\div 95$

Review 4 B
Perform operations with numbers

# 1	$98\times 5\times 4$	# 11	$96\times 6\times 5$
# 2	$96\times 48\div 8$	# 12	$98\times 48 \div 6$
# 3	$30 \div 6\times 908$	# 13	$40 \div 8\times 906$
# 4	$5136 \div 8 \div 6$	# 14	$3240 \div 6 \div 5$
# 5	$975\times 8 + 200$	# 15	$160 + 968\times 5$
# 6	$95\times 94 - 30$	# 16	$95\times 96 - 20$
# 7	$8200 - 86\times 95$	# 17	$8000- 84\times 95$
# 8	$8172\div 9+92$	# 18	$94 + 7248\div 8$
# 9	$8633\div 97 - 19$	# 19	$8352\div 96 - 27$
# 10	$9895 - 9310\div 98$	# 20	$9695 - 9120\div 96$

Summary 5 Chapter 4 Lesson 10
Four Operations with numbers
Test 4 A
Perform operations with numbers

# 1	$9 \times 3 \times 6$	# 11	$9 \times 6 \times 9$
# 2	$9 \times 30 \div 3$	# 12	$9 \times 90 \div 9$
# 3	$30 \div 3 \times 9$	# 13	$90 \div 9 \times 3$
# 4	$1200 \div 30 \div 4$	# 14	$8100 \div 90 \div 9$
# 5	$90 + 990 \times 9$	# 15	$9000 - 70 \times 80$
# 6	$200 + 7200 \div 9$	# 16	$990 - 900 \div 90$
# 7	950×6	# 17	620×9
# 8	$5850 \div 9$	# 18	$5880 \div 6$
# 9	98×21	# 19	65×54
# 10	$3300 \div 55$	# 20	$6930 \div 70$

Summary 5 Chapter 4 Lesson 10
Four operations with numbers
Test 4 B
Perform operations with numbers

# 1	$9 \times 4 \times 7$	# 11	$9 \times 5 \times 8$
# 2	$9 \times 50 \div 5$	# 12	$9 \times 70 \div 7$
# 3	$50 \div 5 \times 7$	# 13	$70 \div 7 \times 5$
# 4	$2500 \div 50 \div 5$	# 14	$4900 \div 70 \div 7$
# 5	$45 + 995 \times 9$	# 15	$9000 - 90 \times 80$
# 6	$100 + 8100 \div 9$	# 16	$1515 - 900 \div 60$
# 7	840×7	# 17	730×8
# 8	$6080 \div 8$	# 18	$6090 \div 7$
# 9	87×32	# 19	76×43
# 10	$4620 \div 60$	# 20	$5720 \div 65$

Summary 5 Chapter 4 Lesson 11
Review 5 A for Final Exam 5 A

 # 1 Subtract

$$\begin{array}{r} 4000 \\ \underline{-\ 2182} \end{array}$$

 # 2 Multiply

$$\begin{array}{r} 678 \\ \underline{\times\ 7} \end{array}$$

 # 3

$$\begin{array}{r} 83 \\ \underline{\times\ 75} \end{array}$$

 # 4 Divide $6030 \div 9$

 # 5 $7332 \div 78$

Evaluate

 # 6 $34 \times 63 \div 7$

 # 7 $1664 \div 8 \times 9$

 # 8 $3312 \div 9 \div 8$

 # 9 $1000 - 75 \times 9$

 # 10 $6304 - 7101 \div 9$

Summary 5 Chapter 4 Lesson 11
Review 5 B for Final Exam 5 B

 # 1 Subtract

$$\begin{array}{r} 5000 \\ \underline{-\ 3292} \end{array}$$

 # 2 Multiply

$$\begin{array}{r} 789 \\ \underline{\times\ 8} \end{array}$$

 # 3

$$\begin{array}{r} 73 \\ \underline{\times\ 65} \end{array}$$

 # 3 Divide $5120 \div 8$

 # 5 $6794 \div 79$

Evaluate

 # 6 $37 \times 72 \div 9$

 # 7 $1648 \div 8 \times 9$

 # 8 $4608 \div 9 \div 8$

 # 9 $1000 - 85 \times 9$

 # 10 $3739 - 6111 \div 9$

Summary 5 Chapter 4 Lesson 12
Final Exam 5 A
 # 1 - # 10 all
Evaluate

 # 1 Subtract $\begin{array}{r} 2000 \\ \underline{-\ 1092} \end{array}$

 # 2 Multiply $\begin{array}{r} 456 \\ \underline{\times\ 7} \end{array}$

 # 3 $\begin{array}{r} 72 \\ \underline{\times\ 65} \end{array}$

 # 4 Divide $4560 \div 8$

 # 5 $8091 \div 87$

 # 6 $27 \times 54 \div 6$

 # 7 $1728 \div 8 \times 9$

 # 8 $3456 \div 9 \div 8$

 # 9 $1000 - 85 \times 9$

 # 10 $6192 - 6102 \div 9$

Summary 5 Chapter 4 Lesson 12
Final Exam 5 B
 # 1 - # 10 all
Evaluate

 # 1 Subtract

$$\begin{array}{r} 3000 \\ \underline{-\ 2191} \end{array}$$

 # 2 Multiply

$$\begin{array}{r} 567 \\ \underline{\times\ 8} \end{array}$$

 # 3

$$\begin{array}{r} 82 \\ \underline{\times\ 75} \end{array}$$

 # 4 Divide $5670 \div 9$

 # 5 $7332 \div 78$

 # 6 $27 \times 72 \div 9$

 # 7 $1458 \div 9 \times 8$

 # 8 $3888 \div 8 \div 9$

 # 9 $1000 - 95 \times 7$

 # 10 $4616 - 4536 \div 8$

Summary 5 Answers Chapter 4

Lesson 1 Review 1 A

	a	b	c		a	b	c
# 1	16	32	48	#11	27	45	63
# 2	32	64	96	#12	54	90	126
# 3	48	96	144	#13	81	135	189
# 4	64	128	192	#14	108	180	252
# 5	80	160	240	#15	135	225	315
# 6	96	192	288	#16	162	270	378
# 7	112	224	336	#17	189	315	441
# 8	128	256	384	#18	216	360	504
# 9	144	288	432	#19	243	405	567
# 10	160	320	480	#20	270	450	630

Lesson 1 Review 1 B

	a	b	c		a	b	c
# 1	8	8	8	#11	9	9	9
# 2	16	16	16	#12	18	18	18
# 3	24	24	24	#13	27	27	27
# 4	32	32	32	#14	36	36	36
# 5	40	40	40	#15	45	45	45
# 6	48	48	48	#16	54	54	54
# 7	56	56	56	#17	63	63	63
# 8	64	64	64	#18	72	72	72
# 9	72	72	72	#19	81	81	81
# 10	80	80	80	#20	90	90	90

Lesson 2 Review 1 A

	a	b	c		a	b	c
# 1	8	8	8	#11	9	9	9
# 2	16	16	16	#12	18	18	18
# 3	24	24	24	#13	27	27	27
# 4	32	32	32	#14	36	36	36
# 5	40	40	40	#15	45	45	45
# 6	48	48	48	#16	54	54	54
# 7	56	56	56	#17	63	63	63
# 8	64	64	64	#18	72	72	72
# 9	72	72	72	#19	81	81	81
# 10	80	80	80	#20	90	90	90

Summary 5 Answers Chapter 4

Lesson 2 Review 1 B

	a	b	c		a	b	c
# 1	1	1	1	#11	1	1	1
# 2	2	2	2	#12	2	2	2
# 3	3	3	3	#13	3	3	3
# 4	4	4	4	#14	4	4	4
# 5	5	5	5	#15	5	5	5
# 6	6	6	6	#16	6	6	6
# 7	7	7	7	#17	7	7	7
# 8	8	8	8	#18	8	8	8
# 9	9	9	9	#19	9	9	9
# 10	10	10	10	#20	10	10	10

Lesson 3 Review 1 A

	a	b		a	b
#1	600	3000	#6	800	4000
#2	2000	4000	#7	2000	5000
#3	3000	4000	#8	2000	6000
#4	4000	5000	#9	3000	7000
#5	5000	6000	#10	4000	9000

Lesson 3 Review 1 B

	a	b		a	b
#11	800	2000	#16	9000	6000
#12	1000	2000	#17	7000	5000
#13	2000	2000	#18	5000	4000
#14	2000	1000	#19	4000	4000
#15	2000	800	#20	3000	9000

Lesson 4 Review 1 A

	a	b		a	b
#1	1400	1000	#6	1100	900
#2	1200	1000	#7	1000	1500
#3	1100	2000	#8	2200	3000
#4	1000	2000	#9	3200	2500
#5	900	3000	#10	1700	1800

Summary 5 Answers Chapter 4

Lesson 4 Review 1 B

	a	b		a	b
# 11	400	10	# 16	950	990
# 12	270	20	# 17	1170	1190
# 13	100	30	# 18	1680	1390
# 14	100	30	# 19	1780	1590
# 15	40	10	# 20	1480	1790

Lesson 6 Review 1 A

#1	525	# 2	530	#3	535	# 4	540
#5	1230	# 6	1236	#7	1242	# 8	1248
#9	2135	# 10	2142	#11	2149	#12	2156
#13	3240	# 14	3248	#15	3256	#16	3264
#17	4545	# 15	4554	#19	4563	#20	4572

Lesson 6 Review 1 B

#1	3025	# 2	3030	#3	3035	# 4	3040
#5	4230	# 6	4236	#7	4242	# 8	4248
#9	5635	# 10	5642	#11	5649	# 12	5656
#13	7340	# 14	7248	# 15	7256	# 16	7264
#17	4581	# 18	5481	#19	6381	# 20	7281

Lesson 6 Review 2 A

#1	105	#2	106	#3	205	#4	206
#5	305	#6	306	#7	405	#8	406
#9	505	#10	506	#11	605	#12	606
#13	705	#14	706	#15	805	#16	806
#17	905	#18	906	#19	509	#20	508

Lesson 6 Review 2 B

#1	107	#2	108	#3	207	#4	208
#5	307	#6	308	#7	407	#8	408
#9	507	#10	508	#11	607	#12	608
#13	707	#14	708	#15	807	#16	808
#17	907	#18	908	#19	507	#20	506

Summary 5 Answers Chapter 4

Lesson 7 Review 1 A

#1	110	#2	240	#3	390	#4	560
#5	750	#6	441	#7	484	#8	529
#9	576	#10	625	#11	390	#12	378
#13	364	#14	348	#15	330	#16	775
#17	768	#18	759	#19	748	#20	735

Lesson 7 Review 1 B

#1	1435	#2	1428	#3	1419	#4	1408
#5	1395	#6	2295	#7	2288	#8	2279
#9	2268	#10	2255	#11	3355	#12	3348
#13	3339	#14	3328	#15	3315	#16	4615
#17	4608	#18	4599	#19	4588	#20	4575

Lesson 8 Review 1 A

#1	10	#2	26	#3	41	#4	61
#5	20	#6	27	#7	42	#8	62
#9	30	#10	28	#11	43	#12	63
#13	40	#14	29	#15	44	#16	64
#17	50	#18	30	#19	45	#20	65

Lesson 8 Review 1 B

#1	81	#2	31	#3	70	#4	90
#5	82	#6	32	#7	69	#8	89
#9	83	#10	33	#11	68	#12	88
#13	84	#14	34	#15	67	#16	87
#17	85	#18	35	#19	66	#20	86

Lesson 9 Review 4 A

#1	1960	#2	480	#3	4930	#4	109	#5	8000
#6	9300	#7	40	#8	1000	#9	70	#10	9500
#11	3840	#12	490	#13	4890	#14	108	#15	5000
#16	9100	#17	20	#18	1000	#19	60	#20	9300

Lesson 9 Review 4 B

#1	1960	#2	576	#3	4540	#4	107	#5	8000
#6	8900	#7	30	#8	1000	#9	70	#10	9800
#11	2880	#12	784	#13	4530	#14	108	#15	5000
#16	9100	#17	20	#18	1000	#19	60	#20	9600

Summary 5 Answers Chapter 4

Lesson 10 Test 4 A

#1	162	#2	90	#3	90	#4	10	#5	9000
#6	1000	#7	5700	#8	650	#9	2058	#10	60
#11	486	#12	90	#13	30	#14	10	#15	3400
#16	980	#17	5580	#18	980	#19	3510	#20	99

Lesson 10 Test 4 B

#1	252	#2	90	#3	70	#4	10	#5	9000
#6	1000	#7	5880	#8	760	#9	2784	#10	77
#11	360	#12	90	#13	50	#14	10	#15	1800
#16	1500	#17	5840	#18	870	#19	3268	#20	88

Lesson 11 Review 5 A

#1	1818	#2	4746	#3	6225	#4	670	#5	94
#6	306	#7	1872	#8	46	#9	325	#10	5515

Lesson 11 Review 5 B

#1	1708	#2	6312	#3	4745	#4	640	#5	86
#6	296	#7	1854	#8	64	#9	235	#10	3060

Lesson 12 Final Exam 5 A

#1	908	#2	3192	#3	4680	#4	570	#5	93
#6	243	#7	1944	#8	48	#9	235	#10	5514

Lesson 12 Final Exam 5 B

#1	809	#2	4536	#3	6150	#4	630	#5	94
#6	216	#7	1296	#8	54	#9	335	#10	4049

Math Help For College

Summary 5

Chapter 5*

Many-digit Numbers

Summary 5 Chapter 5* Contents
Many-digit numbers
Lesson 1 – 14
Basic Problems

1)Work with lesson 1, read rules and examples,
2) Work with the next lessons 2, 3, 4, 5, ...
3) Solve my examples in your notebooks and check.
4) Grading Scale: A =10 points, B=9points, C=8 points, D=7 points, F=0-6 points, if your lesson has 8 problems. You can combine lessons.
5) If you solve 10 (8) problems correctly, you get an A,
6) If have an A, you can move on to next chapter,
7) If your grade is lower than an A, work with the mistakes,
8) When you finish all of the chapters, solve Final Exam in last chapter.

Summary 5 Chapter 5 Lesson 1

Four-digit numbers

Four-digit numbers content four digits: 1000, 1001 ... 9999
Least four-digit number: 1000. Greatest four-digit number: 9999
Digit counts from right to the left: ones, tens, hundreds, thousands
2468 = 2000 + 400 + 60 + 8
One-digit numbers always less than two, three, four-digit numbers
Four-digit numbers always greater than one, two, three-digit numbers
To compare numbers, compare thousands, hundreds, tens, and ones:

$1234 < 1239 \quad (4 < 9) \qquad 1238 > 1230 \ (8 > 0) \qquad 1234=1234$

Properties of Addition and Subtraction

Two replace addends the sum does not change A + B = B + A
2000 + 3000 = 5000 3000 + 2000 = 5000
The sum of a number and zero is the same number
5000 + 0 = 5000 0 + 5000 = 5000 A + 0 = 0 + A = A
Difference any number and zero is the same number
5000 – 0 = 5000 A – 0 = A
The difference two the same number is zero
5000 – 5000 = 0 A – A = 0

10 = 1 + 9 = 2 + 8 = 3 + 7 = 4 + 6 = 5 + 5
11 = 2 + 9 = 3 + 8 = 4 + 7 = 5 + 6
12 = 3 + 9 = 4 + 8 = 5 + 7 = 6 + 6
13 = 4 + 9 = 5 + 8 = 6 + 7
14 = 5 + 9 = 6 + 8 = 7 + 7
15 = 6 + 9 = 7 + 8
16 = 7 + 9 = 8 + 8
17 = 8 + 9
18 = 9 + 9
Adding four-digit numbers:
Add the ones. Add the tens. Add the hundreds. Add the thousands.
Subtracting four- digit numbers:
Subtract the ones. Subtract the tens. Subtract the hundreds. Subtract the thousands.

		1 1 1	8 13 14 16
5678	9678	2686	9 4 5 6
+ 4321	- 5432	+ 1868	- 3 9 8 7
9999	4246	4554	5 4 6 9

Summary 5 Chapter 5 Lesson 2

Five-digit Numbers

Five-digit number content five digits: 10000, 10001 … 99999
Least five-digit number: 10000.
Greatest five-digit number: 99999
Digit counts from right to the left: ones, tens, hundreds, thousands, ten thousand. Twelve thousand four hundred sixty-eight:
12468 = 10000 + 2000 + 400 + 60 + 8
One-digit numbers always less than two, three, four, five-digit numbers
Five-digit numbers always greater than one, two, three, four-digit numbers. To compare number, compare ten thousand, thousands, hundreds, tens, ones 12345 = 12345 12345 < 12349 (5 < 9)
12348 > 12340 (8 > 0)

Properties of addition and subtraction

To replace addends the sum does not change: A + B = B + A
20000 + 30000 = 50000   30000 + 20000 = 50000
Sum any number and zero is the same number
50000 + 0 = 50000   0 + 50000 = 50000   A + 0 = 0 +A = A
Difference any number and zero is the same number
50000 – 0 = 50000   A – 0 = A
The difference two the same number is zero
50000 – 50000 = 0   A – A = 0

Adding five-digit numbers:
Add the ones. Add the tens. Add hundreds.
Add the thousands. Add the ten thousand.

Subtracting five-digit numbers: Subtract the ones. Subtract the tens.
Subtract the hundreds. Subtract the thousands.
Subtract the ten thousand.

		1 1 1	8 13 14 16
56780	96780	26860	9 4 5 6 0
+ 43210	- 54320	+ 18680	- 3 9 8 7 0
99990	42460	45540	5 4 6 90

Summary 5 Chapter 5 Lesson 3

Six-digit Numbers

Six-digit number content six digits 100000, 100001, … 999999
Least six-digit number is 100000; greatest six-digit number is 999999
Digit counts from right to the left - ones, tens, hundreds, thousands, ten thousand, hundred thousand:
123456 = 100000+20000+3000+400+50+6

Compare six-digit numbers

One-digit number always less than two, three, four, five, six- digit number. Six-digit number always greater than five, four, three, two, one- digit number.
To compare number, compare hundred thousand, ten thousand, thousands, hundreds, tens, ones **123456 = 123456**

123456 < 123459 (6 < 9) **123458 > 123450 (8 > 0)**

Properties six-digit numbers: Letters can represent any numbers

To replace addends the sum does not change: A + B = B + A
200000 + 300000 = 500000 **300000 + 200000 = 500000**
The sum of a number and zero is the same number: A + 0 = 0 + A = A
500000 + 0 = 500000 **0 + 500000 = 500000**
Difference any number and zero is the same number
500000 – 0 = 500000 **A – 0 = A**
The difference two the same number is zero
500000 – 500000 = 0 **A – A = 0**

Addition six-digit numbers: Add the ones. Add the tens. Add the hundreds. Add the thousands. Add the ten thousand. Add the hundred thousand. Subtraction six-digit numbers: Subtract the ones. Subtract the tens. Subtract the hundreds. Subtract the thousands. Subtract the ten-thousands. Subtract the hundred-thousand

		1 1 1 1	8 17 16 15 14 13
654321	777777	456789	9 8 7 6 5 3
+123456	-123456	+345678	- 7 8 9 7 8 9
777777	654321	802467	1 9 7 8 6 4

Summary 5 Chapter 5 Lesson 4

Seven-digit numbers

Seven-digit number contain seven digits: 1,000,000 … 9,999,999
The name of seven-digit number read from right to left are millions, hundred-thousands, ten-thousands, thousands, hundreds, tens, ones.

Standard form of numbers is 123, 1234, 12345, 123456.
The expanded form of number
1,234,567 = 1,000,000 + 200,000 + 30,000 + 4,000 + 500 + 60 + 7 =
1,000,000 + 2×100,000 + 3×10,000 + 4×1,000 + 5×100 + 6×10 + 7

Addition seven-digit numbers: Add the ones, add the tens, add the hundreds, and add the thousands, Add the ten-thousands, add the hundred-thousands, and add the millions.

Subtraction seven-digit numbers: Subtract the ones, subtract the tens, subtract the hundreds, Subtract the thousands, subtract the ten-thousands. Subtract the hundred-thousands, subtract the millions.

```
   1 1             10 12
  1020304        6081012
+ 5060708      - 1020304
  -------        -------
  6081012        5060708
```

Many-digit Numbers

Third new period has nine digit and called million 123, 456, 789
The place value of the third period read from right to left are hundred-millions is 1, ten-millions is 2, and millions is 3, hundred-thousand is 4, ten-thousands is 5, thousands is 6, hundreds is 7, tens is 8, and ones is 9

Roman Numbers

I means 1	**V means 5**	**X means 10**
II means 1+1=2	**III means 1+1+1=3**	**IV means 5–1= 4**
VI means 5+1=6	**VII means 5+1+1=7**	**VIII means 5+1+1+1=8**
IX means 10-1=9	**XI means 10+1=11**	**XII means 10+2 =12**
XIV means 10+4=14	**XIX means 10+9=19**	**XXV=10+10+5 = 25**

Summary 5 Chapter 5 Lesson 5

Multiplication Numbers 1

Third operation with numbers is multiplication $\times$ or $\cdot$
Multiplication is the operation of repeated addition
$2\times 3 = 2 + 2 + 2 = 6$ or $2 \cdot 3 = 2 + 2 + 2 = 6$

First number in multiplication called the factor 2
Second number in multiplication called the factor 3
Result of multiplication called the product 6
Factor $\times$ Factor = Product $2 \times 3 = 6$

Properties of multiplication
Two numbers can multiply in any order
The product will be the same $AB = BA$
$2 \times 3 = 2 + 2 + 2 = 6$ $3 \times 2 = 3 + 3 = 6$
The product of a number and zero is zero
$A \times 0 = 0 \times A = 0$ $5 \times 0 = 0$ $0 \times 5 = 0$
The product of a number and one is the same number
$A \times 1 = 1 \times A = A$ $5 \times 1 = 5$ $1 \times 5 = 5$

Factor $\times$ Factor = Product (P) $2\times3 = 6$ (P)

1. First way

$$\begin{array}{r} 25 \\ \underline{\times 6} \\ 30 \\ \underline{+\,120} \\ 150\ (P) \end{array}$$

1) $25 = 20 + 5$
2) $6 \times 5 = 30$
3) $6 \times 20 = 120$
4) $30 + 120 = 150$

2. Second way

$$\begin{array}{r} \text{Top}\ \ 3\ \ \\ 25 \\ \underline{\times 6} \\ 150\ (P) \end{array}$$

1) $6 \times 5 = 30$
2) Put 0 below 6 (P)
3) Put 3 above 2 (T)
4) $6 \times 2 = 12$
5) $12 + 3 = 15$
6) Put 15 ahead 0 (P)

Memorize Rules and Table of Multiplication

$A + 0 = A$ $0 + A = A$ $A - 0 = A$ $A - A = 0$

$A \times 1 = A$ $1 \times A = A$ $A \div 1 = A$ $A \div A = 1$

$A \times 0 = 0$ $0 \times A = 0$ $0 \div A = 0$ $A \div 0$ None

Table of Multiplication and Division

Number 5 from the left column multiply by number 4 from first row to get answer 20

×÷	1	2	3	4	5	6	7	8	9	10
1	1	2	3	4	5	6	7	8	9	10
2	2	4	6	8	10	12	14	16	18	20
3	3	6	9	12	15	18	21	24	27	30
4	4	8	12	16	20	24	28	32	36	40
5	5	10	15	20	25	30	35	40	45	50
6	6	12	18	24	30	36	42	48	54	60
7	7	14	21	28	35	42	49	56	63	70
8	8	16	24	32	40	48	56	64	72	80
9	9	18	27	36	45	54	63	72	81	90
10	10	20	30	40	50	60	70	80	90	100

Summary 5 Chapter 5 Lesson 6
Multiplication two-digit numbers 2
Example 1

Factor × Factor = Product   2 × 3 = 6 (Product)

First way

```
   69        1) 52=50+2
 × 52        2)  69          3)  69
  138 (P1)       ×2              ×50
+3450 (P2)      138 (P1)        3450 (P 2)
 3588 (P)    4) 138 + 3450 = 3588
```

Second way

Example 2

```
Top   4      1)  2 × 9 = 18                7)  5 × 9 = 45
Top   1      2)  Put 8 below 2 (P1)        8)  Put 5 below 3 (P2)
   69        3)  Put 1 above 6 (top)       9)  Put 4 above 1 (top)
 × 52        4)  2 × 6 = 12               10)  5 × 6 = 30
  138 (P1)   5)  12 + 1 = 13              11)  30 + 4 = 34
+3450 (P2)   6)  Put 13 ahead 8 (P1)      12)  Put 34 ahead 5 (P2)
 3588 (P)
```

Example 3

1. First way

```
  525     1) 525 = 500 + 20 + 5
   ×6     2) 6 × 5 = 30
   30     3) 6 × 20 = 120
  120     4) 6 × 500 = 3000
+3000
 3150
```

2. Second way

Example 4

```
Top 1 3    1) 6×5 = 30                 6) Put 5 ahead 0 (P)
525        2) Put 0 below 6 (P)        7) Put 1 ahead 3 (top)
×6         3) Put 3 above 2 (top)      8) 6 × 5 = 30
3150 (P)   4) 6 × 2 = 12               9) 30 + 1 = 31
           5) 12 + 3 = 15             10) Put 31 ahead 5 (P)
```

Summary 5 Chapter 5 Lesson 7

Multiplication Numbers 3

Example 1

$$\begin{array}{r} 9406 \\ \underline{\times \quad 5} \\ 30 \\ 2000 \\ \underline{+\ 45000} \\ 47030 \end{array}$$

1) $9406 = 9000 + 400 + 6$
2) $5\times 6 = \underline{30}$
3) $5\times 400 = \underline{2000}$
4) $5\times 9000 = \underline{45000}$

Example 2

$$\begin{array}{rr} \text{top} & {\scriptstyle 2\ \ 3} \\ & 9406 \\ & \underline{\times \quad 5} \\ & 47030\ (P) \end{array}$$

1) $5 \times 6 = 3\underline{0}$
2) Put $\underline{0}$ below 5 (P)
3) Put 3 above 0 (top)
4) $5 \times 0 = 0$
5) $3 + 0 = \underline{3}$
6) Put $\underline{3}$ ahead 0 (P)
7) $5 \times 4 = 2\underline{0}$
8) Put $\underline{0}$ ahead 3 (P)
9) Put 2 above 9 (top)
10) $5 \times 9 = 45$
11) $45 + 2 = \underline{47}$
12) Put $\underline{47}$ ahead 0 (P)

Example 3

$$\begin{array}{r} 904 \\ \underline{\times 85} \\ 4520 \\ +\underline{72320} \\ 33220 \end{array}$$

1) $85 = 80 + 5$

2) $$\begin{array}{r} 904 \\ \underline{\times 5} \\ 4520 \end{array}$$

3) $$\begin{array}{r} 904 \\ \underline{\times 80} \\ 72320 \end{array}$$

Example 4

$$\begin{array}{r} {\scriptstyle 3} \\ {\scriptstyle 2} \\ 904 \\ \underline{\times 85} \\ 4520 \\ +\underline{72320} \\ 76840 \end{array}$$

1) $85 = 80 + 5$

2) $$\begin{array}{r} {\scriptstyle 2} \\ 904 \\ \underline{\times \ 5} \\ 4520 \end{array}$$

3) $$\begin{array}{r} {\scriptstyle 3} \\ 904 \\ \underline{\times 80} \\ 72320 \end{array}$$

Summary 5 Chapter 5 Lesson 7
Multiplication Numbers
First Way of Multiplication

Example 5

405	1) 106 = 100 + 6	
×106	2) 405	3) 405
2430 (P1)	× 6	×100
+40500 (P2)	2430 (P1)	40500 (P2)
42930 (P)		

Example 6

6304	1) 15 = 10 + 5	
× 15	2) 6304	3) 6304
31520 (P1)	× 5	× 10
+63040 (P2)	31520 (P1)	63040 (P2)
94560 (P)	4) 31520 + 63040 = 94,560 (Product=P)	

Second Way of Multiplication

Example 7

Top 3	1) 6 × 5 = 30	6) Put 3 ahead 0 (P1)
405	2) Put 0 below 6 (P1)	7) 6 × 4 = 24
×106	3) Put 3 above 0 (top)	8) Put 24 ahead 30 (P1)
2430 (P1)	4) 6 × 0 = 0	9) 405 ×100 = 40500 (P2)
+ 40500 (P2)	5) 0 + 3 = 3	10) 2430 + 42930 = 42930 (P)
42930 (P)		

Example 8

Top 1 2		
6304	1) 5 × 4 = 20	8) Put 5 ahead 2 (P1)
× 15	2) Put 0 below 5 (P1)	9) Put 1 above 6 (top)
31520 (P1)	3) Put 2 above 0 (top)	10) 5×6 = 30
+63040 (P2)	4) 5 × 0 = 0	11) 30 + 1=31
94560 (P)	5) 0 + 2 = 2	12) Put 31 ahead 5 (P1)
	6) Put 2 ahead 0 (P1)	13) 6304 ×10 = 63040 (P2)
	7) 5 × 3 = 15	14) 94560 (P)

Summary 5 Chapter 5 Lesson 8

Division Numbers 1

Fourth operation with numbers is division ÷ or :

Division means separated number by some equal groups

First number in division called the dividend 6
Second number in division called the divisor 2
Result of division called the quotient 3
Dividend ÷ divisor = quotient $6 \div 2 = 3$

Properties of Division

Any number divided by one is the same number $5 \div 1 = 5$
Any number divided by the itself is one $5 \div 5 = 1$
Zero divided by any number is zero $0 \div 5 = 0$
Division by zero is not allowed $5 \div 0$ $A \div 0$ none

Formulas:

1) $A \div 1 = A$
2) $A \div A = 1$
3) $0 \div A = 0$
4) $A \div 0$ none

To check multiplication use division
$2 \times 3 = 6$ Check $6 \div 2 = 3$ or $6 \div 3 = 2$
To check division use multiplication and division
$6 \div 2 = 3$ Check $2 \times 3 = 6$ or $6 \div 3 = 2$

Division without remainder R = 0

```
          5 - Quotient
Divisor 7| 35 – Dividend
         -35
           0 – Remainder
```

Division with remainder R > 0

```
          5 - Quotient
Divisor 7| 36 – Dividend
         -35
           1 – Remainder
```

Summary 5 Chapter 5 Lesson 8
Addition and Subtraction of Numbers
Complete the table 1 (Optional)

$5 + 0 = 5$ $0 + 5 = 5$ $5 - 0 = 5$ $5 - 5 = 0$

Number 5 from the left column add number 4 from first row to get answer 9

	+ -	0		2		4		6		8	
#1	1										10
#2										10	
#3	3								10		
#4								10			
#5	5					9	10				
#6						10					
#7	7				10						
#8				10							
#9	9		10								
#10		10									

Complete the table 2
Multiplication and Division of Numbers

$5 \times 1 = 5$ $1 \times 5 = 5$ $5 \times 0 = 0$ $0 \times 5 = 0$

$5 \div 1 = 5$ $5 \div 5 = 1$ $0 \div 5 = 0$ $5 \div 0$ None

Number 5 from the left column multiply by number 4 from first row to get answer 20 (p 384)

	× ÷		2		4		6		8	
#11	1									9
#12									16	
#13	3							21		
#14							24			
#15	5				20	25				
#16					24					
#17	7			21						
#18			16							
#19	9	9								
#20	10									

Summary 5 Chapter 5 Lesson 9
Division Numbers 2

Example 1 5400 – dividend 8 – divisor 685 – quotient

Every time bring down just one digit

<u>675</u>	1) $8 \times \underline{6} = 48$	Put 6 in quotient
8 | 5400	2) $54 - 48 = 6$	
<u>-48</u>	3) Bring down the 8	
60	4) $8 \times \underline{7} = 56$	Put 7 in quotient
<u>- 56</u>	5) $60 - 56 = 4$	
40	6) Bring down the 0	
<u>-40</u>	7) $8 \times \underline{5} = 40$	Put 5 in quotient
0	8) $40 - 40 = 0$	

Example 2 7800 – dividend 8 – divisor 975 – quotient

Every time bring down just one digit

<u>975</u>	1) $8 \times \underline{9} = 72$	Put 9 in quotient
8 | 7800	2) $78 - 72 = 6$	
<u>-72</u>	3) Bring down the 0	
60	4) $8 \times \underline{7} = 56$	Put 7 in quotient
<u>- 64</u>	5) $60 - 56 = 4$	
40	6) Bring down the 0	
<u>-40</u>	7) $8 \times \underline{5} = 40$	Put 5 in quotient
0	8) $40 - 40 = 0$	

Division Numbers

2540 – dividend 5 – divisor 508 – quotient

Example 3 Every time bring down just one digit

<u>508</u>	1) $5 \times \underline{5} = 25$	Put 5 in quotient
5 | 2540	2) $25 - 25 = 0$	
<u>-25</u>	3) Bring down the 4	
04	4) $5 \times \underline{0} = 0$	Put 0 in quotient
<u>- 0</u>	5) $4 - 0 = 4$	
40	6) Bring down the 0	
<u>-40</u>	7) $5 \times \underline{8} = 40$	Put 8 in quotient
0	8) $40 - 40 = 0$	

Summary 5 Chapter 5 Lesson 10
Division Numbers 3

Example 1 3726 – dividend 54 – divisor 69 quotient

Every time bring down just one digit

$37 < 54$ $54 \times 6 = 324$ $54 \times 7 = 378$ $378 > 372$

69	1) $54 \times 6 = 324$	or	3726 ⌊ 54
54 ⟌ 3726	2) $372 - 324 = 48$		-324 69
-324	3) Bring down the 6		486
486	4) $54 \times 9 = 486$		- 486
-486	5) $486 - 486 = 0$		0
0			

Example 2

7608	1)	$5 \times \underline{7} = 35$	Put 7 in quotient
5 ⟌ 38040	2)	$38 - 35 = 3$	
-35	3)	Bring down the 0	
30	4)	$5 \times \underline{6} = 30$	Put 6 in quotient
-30	5)	$30 - 30 = 0$	
04	6)	Bring down the 4	
-0	7)	$5 \times \underline{0} = 0$	Put 0 in quotient
40	8)	$4 - 0 = 4$	
-40	9)	Bring down the 0	
0	10)	$5 \times \underline{8} = 40$	Put 8 in quotient
	11)	$40 - 40 = 0$	

Example 3

507	1)	$86 \times \underline{5} = 430$	Put 5 in quotient
86 ⟌ 43602	2)	$436 - 430 = 6$	
-430	3)	Bring down the 0	
60	4)	$86 \times \underline{0} = 0$	Put 0 in quotient
-0	5)	$60 - 0 = 60$	
602	6)	Bring down the 2	
- 602	7)	$86 \times \underline{7} = 602$	Put 7 in quotient
0	8)	$602 - 602 = 0$	

Summary 5 Chapter 5 Lesson 10
Division Numbers 3

Example 4

6÷2=3   dividend 6   divisors 2   quotients 3(q)

Every time bring down just one digit

```
      708 (q)
905 | 640740
     -6335
        7240
       -7240
```

1) 905 × <u>7</u> = 6335   Put 7 in quotient (q)
2) 6407 – 6335 = 72
3) Bring down the 4
4) 905 × <u>0</u> = 0   Put 0 in quotient (q)
5) 724 – 0 = 724
6) Bring down the 0
7) 905 × <u>8</u> = 7240   Put 8 in quotient (q)
8) 7240 – 7240 = 0

Example 5

Every time bring down just one digit

```
      807 (q)
506 |408342
    -4048
       3542
      -3542
          0
```

1) 506 × <u>8</u> = 4048   Put 8 in quotient (q)
2) 451 – 448 = 3
3) Bring down the 4
4) 506 × <u>0</u> = 0   Put 0 in quotient (q)
5) 354 – 0 = 354
6) Bring down the 2
7) 506 × <u>7</u> = 3542   Put 7 in quotient (q)
8) 3542 – 3542 = 0

Summary 5 Chapter 5 Lesson 11
Order of Operations with Numbers 1

Multiplication of Numbers

To multiply numbers, work in any order

$405 \times 8 \times 9 = 3240 \times 9 = 29160$ $\quad$ $405 \times 8 \times 9 = 405 \times 72 = 29160$

$506 \times 9 \times 8 = 4554 \times 8 = 36432$ $\quad$ $506 \times 9 \times 8 = 506 \times 72 = 36432$

Multiplication and Division of numbers

To multiply and divide numbers work from left to right

$309 \times 81 \div 9 = 25029 \div 9 = 2781$ $\quad$ $409 \times 81 \div 9 = 33129 \div 9 = 3681$

Division and Multiplication of numbers

To divide and multiply numbers work from left to right

$81 \div 9 \times 309 = 9 \times 309 = 2781$ $\quad$ $81 \div 9 \times 409 = 9 \times 409 = 3681$

Division of numbers

To divide numbers, work from left to right

$64000 \div 80 \div 8 = 800 \div 8 = 100$ $\quad$ $81000 \div 90 \div 9 = 900 \div 9 = 100$

Addition and Multiplication with Parentheses

Always first, work in parentheses

$(17 + 19) + 34 = 36 + 34 = 70$ $\quad$ $43 + (18 + 19) = 43 + 37 = 80$

$(25 + 5) \times 8 = 30 \times 8 = 240$ $\quad$ $6 \times (125+25) = 6 \times 150 = 900$

Addition and Subtraction with Parentheses

Always first, work in parentheses

$(700+300) -300=1000-300=700$ $\quad$ $(700-300) + 300 =400+300 =700$

$9000-(500+500) =9000-1000=8000$ $\quad$ $(1200-500) -500=700-500=200$

Multiplication and Division with Parentheses

Always first, work in parentheses

$706 \times (36\div4) =706\times9=6354$ $\quad$ $24028 \div (16\div4) =24028\div4 =6007$

Expressions with Parentheses:

Always first, work in parentheses

$15 \times (807-200) =15\times607=9105$ $\quad$ $(9000+105) \div15 =9105\div15=607$

$(17472-9000) \div 12=8472\div12=706$ $\quad$ $84072\div(10+ 2)=84072\div12=7006$

Do multiplication or division before addition or subtraction

Summary 5 Chapter 5 Lesson 11
Order of Operations with Numbers 1

Addition and Subtraction Numbers

Addition of numbers

To add numbers, add in any order

Example 1 7000 + 35000 + 3000 = 7000 + 38000 = 45000

7000 + 35000 + 3000 = 42000 + 3000 = 45000

7000 + 35000 + 3000=7000 + 3000 + 35000=10000 + 35000=45000

$$\begin{array}{r} 35000 \\ +\ 7000 \\ \underline{3000} \\ 45000 \end{array}$$

Addition and Subtraction of numbers

To add and subtract numbers work from left to right

Example 2 7000 + 35000 - 3000 = 42000 - 3000 = 39000

$$1)\ \begin{array}{r} 35000 \\ \underline{+\ 7000} \\ 42000 \end{array} \qquad 2)\ \begin{array}{r} 42000 \\ \underline{-3000} \\ 39000 \end{array}$$

Subtraction and Addition of numbers

To subtract and add numbers work from left to right

Example 3 35000 – 3000 + 7000 = 32000 + 7000 = 39000

$$1)\ \begin{array}{r} 35000 \\ \underline{-\ 3000} \\ 32000 \end{array} \qquad 2)\ \begin{array}{r} 32000 \\ \underline{+7000} \\ 39000 \end{array}$$

Subtraction of numbers

To subtract numbers, work from left to right

Example 4 35000 – 3000 - 7000 = 32000 - 7000 = 25000

$$1)\ \begin{array}{r} 35000 \\ \underline{-\ 3000} \\ 32000 \end{array} \qquad 2)\ \begin{array}{r} 32000 \\ \underline{-\ 7000} \\ 25000 \end{array}$$

Summary 5 Chapter 5 Lesson 11

Order of Operations with Numbers 1

Multiplication and Addition of numbers

To multiply and add numbers always first multiply then add

Example 5 $5006 \times 5 + 4970 = 30000$

1) $\begin{array}{r} 5006 \\ \times \underline{\quad 5} \\ 25030 \end{array}$ 2) $\begin{array}{r} 25030 \\ \underline{+ 4970} \\ 30000 \end{array}$

Example 6 $1960 + 6005 \times 8 = 50000$

1) $\begin{array}{r} 6005 \\ \times \underline{\quad 8} \\ 48040 \end{array}$ 2) $\begin{array}{r} 48040 \\ \underline{+1960} \\ 50000 \end{array}$

Multiplication and Subtraction of numbers

To multiply and subtract numbers always first multiply then subtract

Example 7 $506 \times 80 - 30480 = 10000$

1) $\begin{array}{r} 506 \\ \underline{\times \ 80} \\ 40480 \end{array}$ 2) $\begin{array}{r} 40480 \\ \underline{- 30480} \\ 10000 \end{array}$

Example 8 $55600 - 608 \times 75 = 10000$

1) $\begin{array}{r} 608 \\ \underline{\times 75} \\ 3040 \\ +\underline{42560} \\ 45600 \end{array}$ 2) $\begin{array}{r} 55600 \\ \underline{- 45600} \\ 10000 \end{array}$

Grade 5 Chapter 5 Lesson 11

Order of Operations with Numbers 1

Division and Addition numbers

To divide and add numbers always first divide then add

Example 9

$$\overset{2}{7696+11520}\overset{1}{\div 5}=10000$$

2304

1) $5\overline{)11520}$ 2) 7696

-10 +2304

15 10000

-15

20

-20

Division and Subtraction numbers

To divide and subtract numbers, always first divide then subtract

Example 10

$$\overset{2}{907-45192}\overset{1}{\div 56}=100$$

807

1) $56\overline{)45192}$ 2) 907

-448 -807

392 100

-392

Summary 5 Chapter 5 Lesson 12

Order of Operations with Numbers 2

Multiplication Numbers

To multiply numbers, work in any order

Example 1 $405 \times 8 \times 9 = 3240 \times 9 = 29160$

1) $\begin{array}{r} 405 \\ \underline{\times 8} \\ 3240 \end{array}$ 2) $\begin{array}{r} 3240 \\ \underline{\times 9} \\ 29160 \end{array}$

Example 2 $405 \times 8 \times 9 = 405 \times 72 = 29160$

1) $\begin{array}{r} 405 \\ \underline{\times 72} \\ 810 \\ \underline{28350} \\ 29160 \end{array}$

Multiplication and Division Numbers

To multiply and divide numbers work from left to right

Example 3 $309 \times 81 \div 9 = 25029 \div 9 = 2781$

1) $\begin{array}{r} 309 \\ \underline{\times 81} \\ 309 \\ \underline{+24720} \\ 25029 \end{array}$

2) $\begin{array}{r} 3681 \\ 9\overline{)25029} \\ \underline{-18} \\ 70 \\ \underline{-63} \\ 72 \\ \underline{-72} \\ 9 \\ \underline{-9} \end{array}$

Example 4 $208 \times 24 \div 6 = 832$

1) $\begin{array}{r} 208 \\ \underline{\times 24} \\ 832 \\ \underline{+4160} \\ 4992 \end{array}$

2) $\begin{array}{r} 832 \\ 6\overline{)4992} \\ \underline{-48} \\ 19 \\ \underline{-18} \\ 12 \\ \underline{-12} \end{array}$

Summary 5 Chapter 5 Lesson 12
Order of Operations with Numbers 2

Multiplication and Addition of numbers

To multiply and add numbers always first multiply then add

Example 5

$$5006 \times 5 + 4970 = 30000$$

1)
$$\begin{array}{r} 5006 \\ \times\ \underline{\quad 5} \\ 25030 \end{array}$$

2)
$$\begin{array}{r} 25030 \\ \underline{+\ 4970} \\ 30000 \end{array}$$

Example 6

$$1960 + 6005 \times 8 = 50000$$

1)
$$\begin{array}{r} 6005 \\ \times\ \underline{\quad 8} \\ 48040 \end{array}$$

2)
$$\begin{array}{r} 48040 \\ \underline{+1960} \\ 50000 \end{array}$$

Multiplication and Subtraction of numbers

To multiply and subtract numbers always first multiply then subtract

Example 7

$$506 \times 80 - 30480 = 10000$$

1)
$$\begin{array}{r} 506 \\ \underline{\times\ 80} \\ 40480 \end{array}$$

2)
$$\begin{array}{r} 40480 \\ \underline{-\ 30480} \\ 10000 \end{array}$$

Example 8

$$55600 - 608 \times 75 = 10000$$

1)
$$\begin{array}{r} 608 \\ \underline{\times 75} \\ 3040 \\ +\underline{42560} \\ 45600 \end{array}$$

2)
$$\begin{array}{r} 55600 \\ \underline{-\ 45600} \\ 10000 \end{array}$$

Summary 5 Chapter 5 Lesson 13
Order of Operations with Numbers 3
Division and Addition of numbers

To divide and add numbers always first divide then add

Example 1 $193230 \div 95 + 7966 = 10000$

1)
```
      2034
95) 193230
   -190
      323
    -285
       380
     -380
```

2) 7966 +2034=10000

Example 2 $7696 + 218880 \div 95 = 10000$

1)
```
      2304
95 )218880
  -190
     288
   -285
       380
      -380
```

2) 7696 +2304=10000

Division and Subtraction of numbers

To divide and subtract numbers always first divide after then subtracts

Example 3 $462060 \div 765 - 104 = 500$

1)
```
          604
765) 462060
    -4590
       3060
      -3060
```

2) 604-104=500

Example 4 $1307 - 462060 \div 765 = 500$

1)
```
        807
706)569742
   -5648
      4942
     -4942
```

2) 1307-807=500

Summary 5 Chapter 5 Lesson 13
Order of Operations with Numbers 3

Expressions with Parentheses

Always first, work in parentheses

Example 5 $15 \times (807 - 200) = 15 \times 607 = 9105$

1)
$$\begin{array}{r} 807 \\ \underline{-200} \\ 607 \end{array}$$

2)
$$\begin{array}{r} 607 \\ \underline{\times 15} \\ 3035 \\ \underline{+6070} \\ 9105 \end{array}$$

Example 6 $(200 + 105) \times 15 = 305 \times 15 = 4575$

1)
$$\begin{array}{r} 200 \\ \underline{+105} \\ 305 \end{array}$$

2)
$$\begin{array}{r} 305 \\ \underline{\times 15} \\ 1525 \\ \underline{+3050} \\ 4575 \end{array}$$

Example 7 $(17472 - 9000) \div 12 = 8472 \div 12 = 706$

1) $17472 - 9000 = 8472$

2)
$$\begin{array}{r} 706 \\ 12\overline{)8472} \\ \underline{-84} \\ 72 \\ \underline{-72} \end{array}$$

Example 8 $84072 \div (10 + 2) = 84072 \div 12 = 7006$

1) $10 + 2 = 12$

2)
$$\begin{array}{r} 7006 \\ 12\overline{)84072} \\ \underline{-84} \\ 072 \\ \underline{-072} \end{array}$$

Summary 5 Chapter 5 Lesson 14
Order of Operations with Numbers 4
Four operations with numbers

To multiply numbers, work in any order

$9 \times 8 \times 7 = 72 \times 7 = 504$ $\quad$ $9 \times 8 \times 7 = 9 \times 56 = 504$

Multiplication and Division Numbers

To multiply and divide numbers work from left to right

$9 \times 6 \div 3 = 54 \div 3 = 18$ $\quad$ $4 \times 81 \div 9 = 324 \div 9 = 36$

Division and Multiplication Numbers

To divide and multiply numbers work from left to right

$24 \div 6 \times 2 = 4 \times 2 = 8$ $\quad$ $81 \div 9 \times 4 = 9 \times 4 = 36$

Division Numbers

To divide numbers, work from left to right

$24 \div 4 \div 2 = 6 \div 2 = 3$ $\quad$ $810 \div 9 \div 3 = 90 \div 3 = 30$

Multiplication and Addition Numbers

To multiply and add numbers always first multiply than add

$9 \times 6 + 9 = 54 + 9 = 63$ $\quad$ $9 + 9 \times 9 = 9 + 81 = 90$

Multiplication and Subtraction Numbers

To multiply and subtract numbers always first multiply than subtract

$9 \times 9 - 9 = 81 - 9 = 72$ $\quad$ $90 - 9 \times 9 = 90 - 81 = 9$

Division and Addition Numbers

To divide and add numbers always first divide than add

$50 \div 5 + 39 = 10 + 39 = 49$ $\quad$ $88 + 171 \div 9 = 88 + 19 = 107$

Division and Subtraction Numbers

To divide and subtract numbers always first divide than subtract

$108 \div 6 - 9 = 18 - 9 = 9$ $\quad$ $500 - 126 \div 7 = 500 - 18 = 482$

Do multiplication or division before addition or subtraction

Signs of operations # 1

Put + or – between numbers that each problem was correct

20 ? 60 = 80 $\quad$ $20 + 60 = 80$ $\quad$ 60 ? 20 = 40 $\quad$ $60 - 20 = 40$

20 ? 70 = 90 $\quad$ $20 + 70 = 90$ $\quad$ 90 ? 20 = 70 $\quad$ $90 - 20 = 70$

Signs of operations # 2

Put × or ÷ between numbers that each problem was correct

6 ? 2 = 12 $\quad$ $6 \times 2 = 12$ $\quad$ 6 ? 2 = 3 $\quad$ $6 \div 2 = 3$

9 ? 3 = 27 $\quad$ $9 \times 3 = 27$ $\quad$ 9 ? 3 = 3 $\quad$ $9 \div 3 = 3$

Summary 5
Chapter 6*

Multiplication Numbers 3

Summary 5 Chapter 6 Contents
Multiplication Numbers
Lesson 1 – 10
Basic Problems

1)Work with lesson 1, read rules and examples,
2) Work with the next lessons 2, 3, 4, 5, …
3) Solve my examples in your notebook and check.
4) Grading Scale: A =10 points, B=9points, C=8 points, D=7 points, F=0-6 points, if your lesson has 8 problems. You can combine lessons.
5) If you solve 10 (8) problems correctly, you get an A,
6) If have an A, you can move on to next chapter,
7) If your grade is lower than an A, work with the mistakes,
8) When you finish all of the chapters, solve Final Exam in last chapter.

Summary 5 Chapter 6 Lesson 1
Multiplication Numbers 1

Third operation with numbers is multiplication × or ·
Multiplication is the operation of repeated addition
$2\times 3 = 2 + 2 + 2 = 6$ or $2 \cdot 3 = 2 + 2 + 2 = 6$

First number in multiplication called the factor 2
Second number in multiplication called the factor 3
Result of multiplication called the product 6
Factor × Factor = Product $2 \times 3 = 6$

Properties of multiplication

Two numbers can multiply in any order
The product will be the same AB = BA
$2 \times 3 = 2 + 2 + 2 = 6$ $3 \times 2 = 3 + 3 = 6$
The product of a number and zero is zero
$A \times 0 = 0 \times A = 0$ $5 \times 0 = 0$ $0 \times 5 = 0$
The product of a number and one is the same number
$A \times 1 = 1 \times A = A$ $5 \times 1 = 5$ $1 \times 5 = 5$

Memorize the Table of Multiplication
Number 5 from the left column multiply by number 4 from first row to get answer 20

×	2	3	<u>4</u>	5	6	7	8	9
1	2	3	4	5	6	7	8	9
2	4	6	8	10	12	14	16	18
3	6	9	12	15	18	21	24	27
4	8	12	16	20	24	28	32	36
<u>5</u>	10	15	<u>20</u>	25	30	35	40	45
6	12	18	24	30	36	42	48	54
7	14	21	28	35	42	49	56	63
8	16	24	32	40	48	56	64	72
9	18	27	36	45	54	63	72	81
10	20	30	40	50	60	70	80	90

Summary 5 Chapter 6 Lesson 1

Multiplication Numbers 1

Factor × Factor = Product $2 \times 3 = 6$

Example 1

1) $15 \times 9 = (10 + 5) \times 9 =$
2) $15 = 10 + 5$
3) $10 \times 9 = 90$ (Product 1 = P 1)
4) $5 \times 9 = 45$ (Product 2 = P 2)
5) $90 + 45 = 135$ (Product = P)
6) $15 \times 9 = 135$

Example 2

$$\begin{array}{r} 1056 \\ \underline{\times \quad 5} \\ 30 \\ 250 \\ + \underline{5000} \\ 5280 \end{array}$$

1) $1056 = 1000 + 50 + 6$
2) $5 \times 6 = \underline{30}$
3) $5 \times 50 = \underline{250}$
4) $5 \times 1000 = \underline{5000}$

Example 3

$$\begin{array}{r} 2056 \\ \underline{\times \quad 5} \\ 30 \\ 250 \\ + \underline{10000} \\ 10280 \end{array}$$

1) $2056 = 2000 + 50 + 6$
2) $5 \times 6 = \underline{30}$
3) $5 \times 50 = \underline{250}$
4) $5 \times 2000 = \underline{10000}$

Summary 5 Chapter 6 Lesson 2
Multiplication numbers 2
Factor × Factor = Product (P)   2×3 = 6 (P)

Example 1

3506	1) 3506 = 3000 + 500 + 6
× 5	2) 5×6 = 30
30	3) 5×500 = 2500
2500	4) 5× 3000 = 15000
+ 15000	
17530	

Example 2

4506	1) 4506 = 4000 + 500 + 6
× 5	2) 5×6 = 30
30	3) 5×500 = 2500
2500	4) 5× 4000 = 20000
+ 20000	
22530	

Example 3

5506	1) 5506 = 5000 + 500 + 6
× 5	2) 5×6 = 30
30	3) 5×500 = 2500
2500	4) 5× 5000 = 25000
+ 25000	
27530	

Example 4

Top 3 2	1) 5 × 4 = 20	7) 5 × 7 = 35
6704	2) Put 0 below 5 (P)	8) Put 5 ahead 2 (P)
× 5	3) Put 2 above 0 (top)	9) Put 3 above 6 (top)
33520 (P)	4) 5 × 0 = 0	10) 5 × 6 = 30
	5) 2 + 0 = 2	11) 30 + 3 = 33
	6) Put 2 ahead 0 (P)	12) Put 33 ahead 5 (P)

Summary 5 Chapter 6 Lesson 3
Multiplication Numbers 3

Factor × Factor = Product (P) **2×3 = 6 (P)**

Example 1

```
Top  2 3
  7406
×    5
 37030 (P)
```

1) 5 × 6 = 3<u>0</u>
2) Put <u>0</u> below 5 (P)
3) Put 3 above 0 (top)
4) 5 × 0 = 0
5) 3 + 0 = <u>3</u>
6) Put <u>3</u> ahead 0 (P)
7) 5 × 4 = 2<u>0</u>
8) Put <u>0</u> ahead 3 (P)
9) Put 2 above 7 (top)
10) 5 × 7 = 35
11) 35 + 2 = <u>37</u>
12) Put <u>37</u> ahead 0 (P)

Example 2

```
Top  2 3
  8406
×    5
 42030 (P)
```

1) 5×6=3<u>0</u>
2) Put <u>0</u> below 5 (P)
3) Put 3 above 0 (top)
4) 5×0=0
5) 3+0=<u>3</u>
6) Put <u>3</u> ahead 0 (P)
7) 5×4=2<u>0</u>
8) Put <u>0</u> in ahead 3 (P)
9) Put 2 above 8 (top)
10) 5×8=40
11) 40+2=<u>42</u>
12) Put <u>42</u> ahead 0 (P)

Example 3

```
Top  2 3
  9406
×    5
 47030 (P)
```

1) 5 × 6 = 3<u>0</u>
2) Put <u>0</u> below 5 (P)
3) Put 3 above 0 (top)
4) 5 × 0 = 0
5) 3 + 0 = <u>3</u>
6) Put <u>3</u> ahead 0 (P)
7) 5 × 4 = 2<u>0</u>
8) Put <u>0</u> ahead 3 (P)
9) Put 2 above 9 (top)
10) 5 × 9 = 45
11) 45 + 2 = <u>47</u>
12) Put <u>47</u> ahead 0 (P)

Summary 5 Chapter 6 Lesson 4
Multiplication Numbers 4

Example 1

$$\begin{array}{r} 104 \\ \times\ 15 \\ \hline 520 \\ +1040 \\ \hline 1560 \end{array}$$

1) 15 = 10 + 5

2) $\begin{array}{r} 104 \\ \times 5 \\ \hline 520 \end{array}$

3) 104 ×10 =1040

Example 2

$$\begin{array}{r} 204 \\ \times\ 15 \\ \hline 1020 \\ +2040 \\ \hline 3060 \end{array}$$

1) 15 = 10 + 5

2) $\begin{array}{r} 204 \\ \times 5 \\ \hline 1020 \end{array}$

3) 204 ×10 =2040

Example 3

$$\begin{array}{r} 304 \\ \times\ 25 \\ \hline 1520 \\ +\ 6080 \\ \hline 7600 \end{array}$$

1) 25 = 20 + 5

2) $\begin{array}{r} 304 \\ \times 5 \\ \hline 1520 \end{array}$

3) $\begin{array}{r} 304 \\ \times 20 \\ \hline 6080 \end{array}$

Example 4

$$\begin{array}{r} 404 \\ \times\ 35 \\ \hline 2020 \\ +12120 \\ \hline 14140 \end{array}$$

1) 35 = 30 + 5

2) $\begin{array}{r} 404 \\ \times 5 \\ \hline 2020 \end{array}$

3) $\begin{array}{r} 404 \\ \times 30 \\ \hline 12120 \end{array}$

Example 5

$$\begin{array}{r} 506 \\ \times 45 \\ \hline 2530 \\ 20240 \\ \hline 22770 \end{array}$$

1) 45 = 40 + 5

2) $\begin{array}{r} 506 \\ \times 5 \\ \hline 2530 \end{array}$

3) $\begin{array}{r} 506 \\ \times 40 \\ \hline 20240 \end{array}$

Summary 5 Chapter6 Lesson 5
Multiplication Numbers 5

Example 1

```
  604    1)  55 = 50 + 5
 × 55    2)  604         3) 604
 3020         ×5            ×50
+30200       3020         30200
 33220
```

Example 2

```
   3
   2                          2                 3
  705   1) 65 = 60 + 5   2) 705       3)     705
  ×65                       × 5              ×60
 3525                      3525            42300
+42300
 45825
```

Example 3

```
   2
   1                                  1           2
  803   1) 75 = 70 + 5     2)       803   3)    803
 × 75                               × 5        × 70
 4015                              4015       56210
+56210
 60225
```

Example 4

```
   3
   2                                  2           3
  904   1) 85 = 80 + 5     2)       904   3)    904
 × 85                               × 5        × 80
 4520                              4520       72320
+72320
 76840
```

Summary 5 Chapter 6 Lesson 6
Multiplication numbers 6
Factor × Factor = Product 2 × 3 = 6

Example 1

Factor × Factor = Product 2 × 3 = 6

```
  14005      1)  14005=10000 + 4000 + 5
×     6      2)  6 × 5 = 30             (Product 1 = P 1)
     30 (P1) 3)  6×4000 = 24000         (Product 2 = P 2)
  24000 (P2) 4)  6×10000 = 60000        (Product 3 = P 3)
  60000 (P3) 5)  60000 + 24000 + 30 = 84030 (Product = P)
  84030 (P)
```

Example 2

```
Top   5  3
  19005
×     6
114030(P)
```

1) 6× 5 = 30
2) Put 0 below 6 (P)
3) Put 3 above 0(top)
4) 6 × 0 = 0
5) 0 + 3 = 3
6) Put 3 ahead 0 (P)
7) 6 × 0 = 0
8) Put 0 ahead 3 (P)
9) 6 × 9 = 54
10) Put 4 ahead 0 (P)
11) Put 5 above 1(top)
12) 6 × 1 = 6
13) 6 + 5 = 11
14) Put 11 ahead 4(P)

Example 3

```
   1004       1)  15 = 10 + 5
 ×   15       2)  1004 ×10 =10040 (Product 2 = P2)
   5020 (P1)  3)  1004
 +10040 (P2)        ×5
  15060 (P)       5020 (Product 1 = P1)
              4)  5020 + 10040 = 15060 (Product = P)
```

Example 4

```
   6304       1) 15 = 10 + 5
 ×   15       2) 6304         3) 6304
  31520 (P1)       × 5             × 10
 +63040 (P2)     31520 (P1)      63040 (P2)
  94560 (P)   4) 31520 +63040 = 94560 (Product=P)
```

Summary 5 Chapter 6 Lesson 7
Multiplication numbers 7

Example 1

405	1) 106 = 100 + 6	
×106	2) 405	3) 405
2430 (P1)	× 6	×100
+40500 (P2)	2430 (P1)	40500 (P2)
42930 (P)		

Example 2

15	2004
×2004	× 15
60	10020
0000	+20040
30000	30060
30060	(easy)

Example 3

16	2005
×2005	× 16
80	12030
0000	+20050
32000	32080
32080	(easy)

Example 4

6304	15
× 15	×6304
31520	60
+63040	4500
94560	90000
(easy)	94560

Example 5

6408	25
× 25	×6408
32040	200
+128160	10000
160200	+150000
(easy)	160200

Summary 5 Chapter 6 Lesson 8

Solve examples yourself then check your solutions

Examples

Multiply Numbers

1) 1056	2) 2056	3) 3506	4) 4506	5) 5506
× 5	× 5	× 5	× 5	× 5
30	30	30	30	30
250	250	2500	2500	2500
+5000	+10000	+15000	+20000	+ 25000
5280	10280	17530	22530	27530

6) 6704	7) 7406	8) 8406	9) 9406	10) 9046
× 5	× 5	× 5	× 5	× 5
20	30	30	30	30
3500	2000	2000	2000	200
+30000	+35000	+40000	+45000	+45000
33520	37030	42030	47030	47230

11) 104	12) 204	13) 304	14) 404	15) 504
×15	×15	×15	×15	×15
520	1020	1520	2020	2520
1040	2040	3040	4040	5040
1560	3060	4560	6060	7560

16) 604	17) 704	18) 804	19) 904	20) 25
×25	×25	×25	×25	× 904
3020	3520	4020	4520	100
+12080	+14080	+16080	+ 18080	+22500
15100	17600	20100	22600	22600

Summary 5 Chapter 6 Lesson 9

Solve examples yourself then check your solutions

Examples

Multiply Numbers

1) 12001	2) 13002	3) 23004	4) 34005	5) 56007
× 9	× 8	× 7	× 6	× 5
108009	104016	161028	204030	280035

6) 67004	7) 34005	8) 45006	9) 56007	10) 67008
× 9	× 8	× 7	× 6	× 5
603036	272040	315042	336042	335040

11) 604	12) 704	13) 804	14) 904	15) 205
×205	×205	× 205	×205	× 904
3020	3520	4020	4520	820
+120800	+140800	+160800	+180800	+184500
123820	144320	164820	185320	185320

16) 1200	17) 2304	18) 3405	19) 4506	20) 5607
× 15	× 25	× 35	×45	× 65
6000	11520	17025	22530	28035
+12000	+46080	+102150	+ 180240	+336420
18000	57600	119175	202770	364455

Summary 5 Chapter 6 Lesson 10

1) $\begin{array}{r} 607 \\ \times 708 \\ \hline 4856 \\ +424900 \\ \hline 429756 \end{array}$

2) $\begin{array}{r} 708 \\ \times 809 \\ \hline 6372 \\ +566400 \\ \hline 572772 \end{array}$

3) $\begin{array}{r} 809 \\ \times 908 \\ \hline 6472 \\ +728100 \\ \hline 734572 \end{array}$

4) $\begin{array}{r} 908 \\ \times 807 \\ \hline 6356 \\ +726400 \\ \hline 732756 \end{array}$

5) $\begin{array}{r} 807 \\ \times 706 \\ \hline 4842 \\ +564900 \\ \hline 569742 \end{array}$

<u>Summary 5</u>
Chapter 7*

Division
Numbers 3

Summary 5 Chapter 7 Contents
Division Numbers, Lesson 1 – 10
Basic Problems

1)Work with lesson 1, read rules and examples,
2) Work with the next lessons 2, 3, 4, 5, ...
3) Solve my examples in your notebook and check.
4) Grading Scale: A =10 points, B=9points, C=8 points, D=7 points, F=0-6 points, if your lesson has 8 problems. You can combine lessons.
5) If you solve 10 (8) problems correctly, you get an A,
6) If have an A, you can move on to next chapter,
7) If your grade is lower than an A, work with the mistakes,
8) When you finish all of the chapters, solve Final Exam in last chapter.

Summary 5 Chapter 7 Lesson 1
Division Numbers 1

Fourth operation with numbers is division ÷ or :
Division means separated number by some equal groups

First number in division called the dividend	**6**
Second number in division called the divisor	**2**
Result of division called the quotient	**3**
Dividend ÷ divisor = quotient	**6÷2 = 3**

Properties of division

Any number divided by one is the same number	**5 ÷ 1 = 5**
Any number divided by the itself is one	**5 ÷ 5 = 1**
Zero divided by any number is zero	**0 ÷ 5 = 0**
Division by zero is not allowed	**5 ÷ 0 A ÷ 0 None**

Formulas: **1)** $A \div 1 = A$ **2)** $A \div A = 1$
3) $0 \div A = 0$ **4)** $A \div 0$ **None**

Review 1 A
Divide numbers by using Table of multiplication (p 384)

	a	b	c	d
1)	0 ÷ 6	7 ÷ 7	72 ÷ 8	0 ÷ 9
2)	12 ÷ 6	21 ÷ 7	64 ÷ 8	63 ÷ 9
3)	30 ÷ 6	28 ÷ 7	48 ÷ 8	45 ÷ 9
4)	42 ÷ 6	42 ÷ 7	32 ÷ 8	27 ÷ 9
5)	48 ÷ 6	63 ÷ 7	16 ÷ 8	9 ÷ 9
6)	54 ÷ 6	56 ÷ 7	8 ÷ 8	18 ÷ 9
7)	36 ÷ 6	49 ÷ 7	24 ÷ 8	36 ÷ 9
8)	24 ÷ 6	35 ÷ 7	40 ÷ 8	54 ÷ 9
9)	18 ÷ 6	14 ÷ 7	56 ÷ 8	72 ÷ 9
10)	6 ÷ 6	0 ÷ 7	0 ÷ 8	81 ÷ 9

Summary 5 Chapter 7 Lesson 1
Division Numbers 1

To check multiplication use division

2×3=6 Check **6÷2=3 or 6÷3=2**

To check division use multiplication and division

6÷2=3 Check 2×3=6 or 6÷3=2

Division without remainder R = 0

```
              5 - Quotient
Divisor 7⌋ 35 – Dividend
          -35
            0 – Remainder
```

Division with remainder R > 0

```
              5 - Quotient
Divisor 7 |36 – Dividend
          -35
            1 – Remainder
```

Review 1 B

Complete the Table of Division by using Table of Multiplication (p 384)

		a	b	c	d	e	f	g	h
	×	2	?	4	?	6	?	8	?
11)	1								9
12)	?								18
13)	3							24	
14)	?							32	
15)	5						35		
16)	?					36			
17)	7				35				
18)	?			32					
19)	9		27						
20)	?	20							

Summary 5 Chapter 7 Lesson 2

Division Numbers 2

6÷2=3 dividend 6 divisor 2 quotient 3(q)

Every time bring down just one digit

Example 1

105 (q)	1)	$14 \times \underline{1} = 14$	Put 1 in quotient (q)
14 | 1470	2)	$14 - 14 = 0$	
-10	3)	Bring down the 7	
07	4)	$14 \times \underline{0} = 0$	Put 0 in quotient (q)
-0	5)	$7 - 0 = 7$	
70	6)	Bring down the 0	
- 70	7)	$14 \times \underline{5} = 70$	Put 5 in quotient(q)
0	8)	$70 - 70 = 0$	

Example 2

125 (q)	1)	$18 \times \underline{1} = 18$	Put 1 in quotient (q)
18 | 2250	2)	$22 - 18 = 4$	
-18	3)	Bring down the 5	
45	4)	$18 \times \underline{2} = 36$	Put 2 in quotient (q)
-36	5)	$45 - 36 = 9$	
90	6)	Bring down the 0	
-90	7)	$18 \times \underline{5} = 90$	Put 5 in quotient (q)
0	8)	$90 - 90 = 0$	

Summary 5 Chapter 7 Lesson 2

Division Numbers 2

$6 \div 2 = 3$ dividend 6 divisor 2 quotient 3 (q)

Every time bring down just one digit

Example 3

$\underline{206}$ (q)	1)	$5 \times \underline{2} = 10$	or 1030⌊8
5 | 1030	2)	$10 - 10 = 0$	$-\underline{10}$ 206 (q)
$\underline{-10}$	3)	Bring down the 3	03
03	4)	$5 \times \underline{0} = 0$	$\underline{-0}$
$\underline{-0}$	5)	$3 - 0 = 3$	30
30	6)	Bring down the 0	$\underline{-30}$
$\underline{-30}$	7)	$5 \times \underline{6} = 30$	0
0	8)	$30 - 30 = 0$	

Example 4

$\underline{295}$ (q)	1)	$8 \times \underline{2} = 16$	or 2360⌊8
8 | 2360	2)	$23 - 16 = 7$	$-\underline{16}$ 295 (q)
$\underline{-16}$	3)	Bring down the 6	76
76	4)	$8 \times \underline{9} = 72$	$\underline{-72}$
$\underline{-72}$	5)	$76 - 72 = 4$	40
40	6)	Bring down the 0	$\underline{-40}$
$\underline{-40}$	7)	$8 \times \underline{5} = 40$	0
0	8)	$40 - 40 = 0$	

Example 5

$\underline{305}$ (q)	1)	$5 \times \underline{3} = 15$	Put 3 in quotient (q)
5 | 1525	2)	$15 - 15 = 0$	
$\underline{-15}$	3)	Bring down the 2	
02	4)	$5 \times \underline{0} = 0$	Put 0 in quotient (q)
$\underline{-0}$	5)	$2 - 0 = 2$	
25	6)	Bring down the 5	
$\underline{-25}$	7)	$5 \times \underline{5} = 25$	Put 5 in quotient (q)
0	8)	$25 - 25 = 0$	

Summary 5 Chapter 7 Lesson 3
Division Numbers 3

6÷2=3 dividend 6 divisor 2 quotient 3 (q)

Every time bring down just one digit

Example 1

```
  395 (q)
8|3160
 -24
  76
 - 72
   40
  -40
    0
```

1) $8 \times \underline{3} = 24$ Put 3 in quotient (q)
2) $31 - 24 = 7$
3) Bring down the 6
4) $8 \times \underline{9} = 72$ Put 9 in quotient (q)
5) $76 - 72 = 4$
6) Bring down the 0
7) $8 \times \underline{5} = 40$ Put 5 in quotient (q)
8) $40 - 40 = 0$

Example 2

```
  407 (q)
5|2035
 -20
  03
  - 0
   35
  -35
    0
```

1) $5 \times \underline{4} = 20$ Put 4 in quotient (q)
2) $20 - 20 = 0$
3) Bring down the 3
4) $5 \times \underline{0} = 0$ Put 0 in quotient (q)
5) $3 - 0 = 3$
6) Bring down the 5
7) $5 \times \underline{7} = 35$ Put 7 in quotient (q)
8) $35 - 35 = 0$

Example 3

```
  495
8|3960
 -32
  76
 - 72
   40
  -40
    0
```

1) $8 \times \underline{4} = 32$ Put 4 in quotient (q)
2) $39 - 32 = 7$
3) Bring down the 6
4) $8 \times \underline{9} = 72$ Put 9 in quotient (q)
5) $76 - 72 = 4$
6) Bring down the 0
7) $8 \times \underline{5} = 40$ Put 5 in quotient (q)
8) $40 - 40 = 0$

Summary 5 Chapter 7 Lesson 3
Division Numbers 3

6÷2=3 dividend 6 divisor 2 quotient 3 (q)

Every time bring down just one digit

Example 4

508 (q)	1)	5 × 5 = 25	Put 5 in quotient (q)
5 ⟌ 2540	2)	25 – 25 = 0	
-25	3)	Bring down the 4	
04	4)	5 × 0 = 0	Put 0 in quotient (q)
- 0	5)	4 – 0 = 4	
40	6)	Bring down the 0	
-40	7)	5 × 8 = 40	Put 8 in quotient (q)
0	8)	40 – 40 = 0	

Example 5

595 (q)	1)	8 × 5 = 40	Put 5 in quotient (q)
8 ⟌ 4760	2)	47 – 40 = 7	
-40	3)	Bring down the 6	
76	4)	8 × 9 = 72	Put 9 in quotient (q)
- 72	5)	76 – 72 = 4	
40	6)	Bring down the 0	
-40	7)	8 × 5 = 40	Put 5 in quotient (q)
0	8)	40 – 40 = 0	

Summary 5 Chapter 7 Lesson 4
Division Numbers 4
Example 1

6÷2=3 dividend 6 divisor 2 quotient 3 (q)

Every time bring down just one digit. Example 1

1056 (q)	1) 5 × 1 = 5	or	5280 | 5	
5 | 5280	2) 5 – 5 = 0		-5	1056 (q)
-5	3) Bring down the 2		02	
02	4) 5 × 0 = 0		-0	
-0	5) 2 – 0 = 2		28	
28	6) Bring down the 8		-25	
-25	7) 5×5 = 25		30	
30	8) 28 – 25 = 3		- 30	
-30	9) Bring down the 0		0	
0	10) 5 × 6 = 30	11)	30 – 30 = 0	

Example 2

1506 (q)	1) 5 × 1 = 5	or	7530 | 5	
5 | 7530	2) 7 – 5 = 2		-5	1506 (q)
-5	3) Bring down the 5		25	
25	4) 5 × 5 = 25		-25	
-25	5) 25 – 25 = 0		03	
03	6) Bring down the 3		-0	
-0	7) 5×0 = 0		30	
30	8) 3 – 0 = 3		-30	
-30	9) Bring down the 0		0	
0	10) 5 × 6 = 30	11)	30 – 30 = 0	

Example 3

6÷2=3 dividend 6 divisor 2 quotient 3 (q)

Every time bring down just one digit. Example 1

2046	1)	5 × 2 = 10	Put 2 in quotient (q)
5 | 10230	2)	10 – 10 = 0	
-10	3)	Bring down the 2	
02	4)	5 × 0 = 0	Put 0 in quotient (q)
-0	5)	2 – 0 = 2	
23	6)	Bring down the 3	
-20	7)	5 × 4 = 20	Put 4 in quotient (q)
30	8)	23 – 20 = 3	
- 30	9)	Bring down the 0	
0	10)	5 × 6 = 30	Put 6 in quotient (q) 11) 30–30=0

Summary 5 Chapter 7 Lesson 5
Division Numbers 4

15280 – dividend 5 – divisor 3056 – quotient

Every time bring down just one digit.

Example 1

3056	1) $5 \times \underline{3} = 15$	Put 3 in quotient
5 ⌈ 15280	2) $15 - 15 = 0$	
-15	3) Bring down the 2	
02	4) $5 \times \underline{0} = 0$	Put 0 in quotient
-0	5) $2 - 0 = 2$	
28	6) Bring down the 8	
-25	7) $5 \times \underline{5} = 25$	Put 5 in quotient
30	8) $28 - 25 = 3$	
-30	9) Bring down the 0	
- 30	10) $5 \times \underline{6} = 30$	Put 6 in quotient
0	11) $30 - 30 = 0$	

Example 2

17530 – dividend 5 – divisor 3506 – quotient

3506	1) $5 \times \underline{3} = 15$	Put 3 in quotient
5 ⌈ 17530	2) $17 - 15 = 2$	
-15	3) Bring down the 5	
25	4) $5 \times \underline{5} = 25$	Put 5 in quotient
-25	5) $25 - 25 = 0$	
03	6) Bring down the 3	
-0	7) $5 \times \underline{0} = 0$	Put 0 in quotient
30	8) $3 - 0 = 3$	
- 30	9) Bring down the 0	
0	10) $5 \times \underline{6} = 30$	Put 6 in quotient
	11) $30 - 30 = 0$	

Summary 5 Chapter 7 Lesson 5
Division Numbers 5

Example 3

20280 – dividend   5 – divisor   4056 – quotient

Every time bring down just one digit.

4056	1)	$5 \times \underline{4} = 20$	Put 4 in quotient
5 ⟌ 20280	2)	$20 - 20 = 0$	
$\underline{-20}$	3)	Bring down the 2	
02	4)	$5 \times \underline{0} = 0$	Put 0 in quotient
$\underline{-0}$	5)	$2 - 0 = 2$	
28	6)	Bring down the 8	
$\underline{-25}$	7)	$5 \times \underline{5} = 25$	Put 5 in quotient
30	8)	$28 - 25 = 3$	
$- \underline{30}$	9)	Bring down the 0	
0	10)	$5 \times \underline{6} = 30$	Put 6 in quotient
	11)	$30 - 30 = 0$	

Example 4

22530 – dividend   5 – divisor   4506 – quotient

Every time bring down just one digit

4506	1)	$5 \times \underline{4} = 20$	Put 4 in quotient
5 ⟌ 22530	2)	$22 - 20 = 2$	
$\underline{-20}$	3)	Bring down the 5	
25	4)	$5 \times \underline{5} = 25$	Put 5 in quotient
$\underline{-25}$	5)	$25 - 25 = 0$	
03	6)	Bring down the 3	
$\underline{-0}$	7)	$5 \times \underline{0} = 0$	Put 0 in quotient
30	8)	$3 - 0 = 3$	
$- \underline{30}$	9)	Bring down the 0	
0	10)	$5 \times \underline{6} = 30$	Put 6 in quotient
	11)	$30 - 30 = 0$	

Summary 5 Chapter 7 Lesson 6
Division Numbers 6
Example 1

25280 – dividend 5 – divisor 5056 – quotient

Every time bring down just one digit

```
    5056
5 | 25280
  -25
    02
    -0
     28
    -25
      30
    - 30
       0
```

1) $5 \times \underline{5} = 25$ Put 5 in quotient
2) $20 - 20 = 0$
3) Bring down the 2
4) $5 \times \underline{0} = 0$ Put 0 in quotient
5) $2 - 0 = 2$
6) Bring down the 8
7) $5 \times \underline{5} = 25$ Put 5 in quotient
8) $28 - 25 = 3$
9) Bring down the 0
10) $5 \times \underline{6} = 30$ Put 6 in quotient
11) $30 - 30 = 0$

Example 2

27530 – dividend 5 – divisor 5506 – quotient

Every time bring down just one digit

```
    5506
5 | 27530
  -25
    25
   -25
    03
    -0
     30
   - 30
      0
```

1) $5 \times \underline{5} = 25$ Put 5 in quotient
2) $27 - 25 = 2$
3) Bring down the 5
4) $5 \times \underline{5} = 25$ Put 5 in quotient
5) $25 - 25 = 0$
6) Bring down the 3
7) $5 \times \underline{0} = 0$ Put 0 in quotient
8) $3 - 0 = 3$
9) Bring down the 0
10) $5 \times \underline{6} = 30$ Put 6 in quotient
11) $30 - 30 = 0$

Summary 5 Chapter 7 Lesson 6
Division Numbers 6
Example 3

30280 – dividend 5 – divisor 6056 – quotient
Every time bring down just one digit

6056	1) 5 × 6 = 30 Put 6 in quotient
5 | 30280	2) 30 – 30 = 0
-30	3) Bring down the 2
02	4) 5 × 0 = 0 Put 0 in quotient
-0	5) 2 – 0 = 2
28	6) Bring down the 8
-25	7) 5 × 5 = 25 Put 5 in quotient
30	8) 28 – 25 = 3
- 30	9) Bring down the 0
0	10) 5 × 6 = 30 Put 6 in quotient
	11) 30 – 30 = 0

Example 4

32530 – dividend 5 – divisor 6506 – quotient
Every time bring down just one digit

6506	1) 5 × 6 = 30 Put 6 in quotient
5 | 32530	2) 32 – 30 = 2
-30	3) Bring down the 5
25	4) 5 × 5 = 25 Put 5 in quotient
-25	5) 25 – 25 = 0
03	6) Bring down the 3
-0	7) 5 × 0 = 0 Put 0 in quotient
30	8) 3 – 0 = 3
- 30	9) Bring down the 0
0	10) 5 × 6 = 30 Put 6 in quotient
	11) 30 – 30 = 0

Summary 5 Chapter 7 Lesson 7
Division Numbers 7
Example 1

35280 – dividend   5 – divisor   7056 – quotient

Every time bring down just one digit

7056	1)	$5 \times \underline{7} = 35$	Put 7 in quotient
5 ∣ 35280	2)	$35 - 35 = 0$	
-35	3)	Bring down the 2	
02	4)	$5 \times \underline{0} = 0$	Put 0 in quotient
-0	5)	$2 - 0 = 2$	
28	6)	Bring down the 8	
-25	7)	$5 \times \underline{5} = 25$	Put 5 in quotient
30	8)	$28 - 25 = 3$	
- 30	9)	Bring down the 0	
0	10)	$5 \times \underline{6} = 30$	Put 6 in quotient
	11)	$30 - 30 = 0$	

Example 2

37530 – dividend   5 – divisor   7506 – quotient

Every time bring down just one digit

7506	1)	$5 \times \underline{7} = 35$	Put 7 in quotient
5 ∣ 37530	2)	$37 - 35 = 2$	
-35	3)	Bring down the 5	
25	4)	$5 \times \underline{5} = 25$	Put 5 in quotient
-25	5)	$25 - 25 = 0$	
03	6)	Bring down the 3	
-0	7)	$5 \times \underline{0} = 0$	Put 0 in quotient
30	8)	$3 - 0 = 3$	
- 30	9)	Bring down the 0	
0	10)	$5 \times \underline{6} = 30$	Put 6 in quotient
	11)	$30 - 30 = 0$	

Summary 5 Chapter 7 Lesson 7
Division Numbers 7
Example 3

40280 – dividend 5 - divisor 8056 – quotient
Every time bring down just one digit

8056	1)	$5 \times \underline{8} = 40$	Put 8 in quotient
5 ⟌ 40280	2)	$40 - 40 = 0$	
-40	3)	Bring down the 2	
02	4)	$5 \times \underline{0} = 0$	Put 0 in quotient
-0	5)	$2 - 0 = 2$	
28	6)	Bring down the 8	
-25	7)	$5 \times \underline{5} = 25$	Put 5 in quotient
30	8)	$28 - 25 = 3$	
- 30	9)	Bring down the 0	
0	10)	$5 \times \underline{6} = 30$	Put 6 in quotient
	11)	$30 - 30 = 0$	

Example 4

42530 – dividend 5 - divisor 8506 – quotient
Every time bring down just one digit
Example 4

8506	1)	$5 \times \underline{8} = 40$	Put 8 in quotient
5 ⟌ 42530	2)	$42 - 40 = 2$	
-40	3)	Bring down the 5	
25	4)	$5 \times \underline{5} = 25$	Put 5 in quotient
-25	5)	$25 - 25 = 0$	
03	6)	Bring down the 3	
-0	7)	$5 \times \underline{0} = 0$	Put 0 in quotient
30	8)	$3 - 0 = 3$	
- 30	9)	Bring down the 0	
0	10)	$5 \times \underline{6} = 30$	Put 6 in quotient
	11)	$30 - 30 = 0$	

Summary 5 Chapter 7 Lesson 8
Division Numbers 8
Example 1

45280 – dividend **5 – divisor** **9056 – quotient**

Every time bring down just one digit

9056	1)	$5 \times \underline{9} = 45$	Put 9 in quotient
5 | 45280	2)	$45 - 45 = 0$	
-45	3)	Bring down the 2	
02	4)	$5 \times \underline{0} = 0$	Put 0 in quotient
-0	5)	$2 - 0 = 2$	
28	6)	Bring down the 8	
-25	7)	$5 \times \underline{5} = 25$	Put 5 in quotient
30	8)	$28 - 25 = 3$	
- 30	9)	Bring down the 0	
	10)	$5 \times \underline{6} = 30$	Put 6 in quotient
	11)	$30 - 30 = 0$	

Example 2

47530 – dividend **5 – divisor** **9506 – quotient**

Every time bring down just one digit

9506	1)	$5 \times \underline{9} = 45$	Put 9 in quotient
5 | 47530	2)	$47 - 45 = 2$	
-45	3)	Bring down the 5	
25	4)	$5 \times \underline{5} = 25$	Put 5 in quotient
-25	5)	$25 - 25 = 0$	
03	6)	Bring down the 3	
-0	7)	$5 \times \underline{0} = 0$	Put 0 in quotient
30	8)	$3 - 0 = 3$	
- 30	9)	Bring down the 0	
0	10)	$5 \times \underline{6} = 30$	Put 6 in quotient
	11)	$30 - 30 = 0$	

Summary 5 Chapter 7 Lesson 8
Division Numbers 8

Example 3
Every time bring down just one digit

105	1)	$15 \times \underline{1} = 15$	Put 1 in quotient
15 | 1575	2)	$15 - 15 = 0$	
-15	3)	Bring down the 7	
07	4)	$15 \times \underline{0} = 0$	Put 0 in quotient
-0	5)	$7 - 0 = 7$	
75	6)	Bring down the 5	
- 75	7)	$15 \times \underline{5} = 75$	Put 5 in quotient
0	8)	$75 - 75 = 0$	

Example 4

Every time bring down just one digit

105	1)	$18 \times \underline{1} = 18$	Put 1 in quotient
18 | 1890	2)	$18 - 18 = 0$	
-18	3)	Bring down the 9	
09	4)	$18 \times \underline{0} = 0$	Put 0 in quotient
-0	5)	$9 - 0 = 9$	
90	6)	Bring down the 0	
-90	7)	$18 \times \underline{5} = 90$	Put 5 in quotient
0	8)	$90 - 90 = 0$	

Summary 5 Chapter 7 Lesson 9
Division Numbers 9
Example 1

Every time bring down just one digit

205	1)	$15 \times \underline{2} = 30$	Put 2 in quotient
15 | 3075	2)	$30 - 30 = 0$	
-30	3)	Bring down the 7	
07	4)	$15 \times \underline{0} = 0$	Put 0 in quotient
-0	5)	$7 - 0 = 7$	
75	6)	Bring down the 5	
- 75	7)	$15 \times \underline{5} = 75$	Put 5 in quotient
0	8)	$75 - 75 = 0$	

Example 2

205	1)	$18 \times \underline{2} = 36$	Put 2 in quotient
18 | 3690	2)	$36 - 36 = 0$	
-36	3)	Bring down the 9	
09	4)	$18 \times \underline{0} = 0$	Put 0 in quotient
-0	5)	$9 - 0 = 9$	
90	6)	Bring down the 0	
-90	7)	$18 \times \underline{5} = 90$	Put 5 in quotient
0	8)	$90 - 90 = 0$	

Example 3

305	1)	$25 \times \underline{3} = 75$	Put 3 in quotient
25 | 7625	2)	$76 - 75 = 1$	
-75	3)	Bring down the 2	
12	4)	$25 \times \underline{0} = 0$	Put 0 in quotient
-0	5)	$12 - 0 = 12$	
125	6)	Bring down the 5	
-125	7)	$25 \times \underline{5} = 125$	Put 5 in quotient
0	8)	$125 - 125 = 0$	

Summary 5 Chapter 7 Lesson 10
Division Numbers 10
Every time bring down just one digit
Example 1

305	1)	$28 \times \underline{3} = 84$	Put 3 in quotient
28 ⟌ 8540	2)	$85 - 84 = 1$	
-84	3)	Bring down the 4	
14	4)	$28 \times \underline{0} = 0$	Put 0 in quotient
-0	5)	$14 - 0 = 14$	
140	6)	Bring down the 0	
-140	7)	$28 \times \underline{5} = 140$	Put 5 in quotient
0	8)	$140 - 140 = 0$	

Summary 5

Chapter 8* Four Operations With Numbers

Summary 5 Chapter 8 Contents
Four Operations with Numbers
Lesson 1 – 12
Basic Problems

1)Work with lesson 1, read rules and examples,

2) Work with the next lessons 2, 3, 4, 5, …

3) Solve my examples in your notebook and check.

4) Grading Scale: A =10 points, B=9points, C=8 points, D=7 points, F=0-6 points, if your lesson has 8 problems. You can combine lessons.

5) If you solve 10 (8) problems correctly, you get an A,

6) If have an A, you can move on to next chapter,

7) If your grade is lower than an A, work with the mistakes,

8) When you finish all of the chapters, solve Final Exam in last chapter.

Summary 5 Chapter 8 Lesson 1

Addition Numbers

To add numbers, add in any order

Examples 1

$$7000 + 35000 + 3000 = 7000 + 38000 = 45000$$
$$7000 + 35000 + 3000 = 42000 + 3000 = 45000$$
$$7000 + 35000 + 3000 = 7000 + 3000 + 35000 = 10000 + 35000 = 45000$$

Add in Column

$$\begin{array}{r} 35000 \\ +\ 7000 \\ \underline{3000} \\ 45000 \end{array}$$

Addition and Subtraction numbers

To add and subtract numbers work from left to right

Example 2

$$7000 + 35000 - 3000 = 42000 - 3000 = 39000$$

Add in Column

$$7000 + 35000 - 3000 = 39000$$

1) $\begin{array}{r} 35000 \\ \underline{+\ 7000} \\ 42000 \end{array}$ 2) $\begin{array}{r} 42000 \\ \underline{-3000} \\ 39000 \end{array}$

Summary 5 Chapter 8 Lesson 2
Subtraction and Addition numbers
To subtract and add numbers work from left to right
Example 1

$35000 - 3000 + 7000 = 32000 + 7000 = 39000$

Add in Column

$35000 - 3000 + 7000 = 32000 + 7000 = 39000$

1) $\begin{array}{r} 35000 \\ -\ 3000 \\ \hline 32000 \end{array}$ 2) $\begin{array}{r} 32000 \\ +7000 \\ \hline 39000 \end{array}$

Subtraction numbers
To subtract numbers, work from left to right
Example 2

$35000 - 3000 - 7000 = 32000 - 7000 = 25000$

Subtraction numbers in column
Add in Column

$35000 - 3000 - 7000 = 32000 - 7000 = 25000$

1) $\begin{array}{r} 35000 \\ -\ 3000 \\ \hline 32000 \end{array}$ 2) $\begin{array}{r} 32000 \\ -\ 7000 \\ \hline 25000 \end{array}$

Summary 5 Chapter 8 Lesson 3
Multiplication numbers
To multiply numbers, work in any order
Examples 1

a) $405 \times 8 \times 9 = 3240 \times 9 = 29160$
b) $405 \times 8 \times 9 = 405 \times 72 = 29160$
c) $506 \times 9 \times 8 = 4554 \times 8 = 36432$
d) $506 \times 9 \times 8 = 506 \times 72 = 36432$

Multiplication and Division numbers
To multiply and divide numbers work from left to right
Examples 2

a) $309 \times 81 \div 9 = 25029 \div 9 = 2781$
b) $409 \times 81 \div 9 = 33129 \div 9 = 3681$

Summary 5 Chapter 8 Lesson 4

Division and Multiplication numbers

To divide and multiply numbers work from left to right

Examples 1

a) $81 \div 9 \times 309 = 9 \times 309 = 2781$

b) $81 \div 9 \times 409 = 9 \times 409 = 3681$

Division numbers

To divide numbers, work from left to right

Examples 2

a) $64000 \div 80 \div 8 = 800 \div 8 = 100$

b) $81000 \div 90 \div 9 = 900 \div 9 = 100$

Summary 5 Chapter 8 Lesson 5

Multiplication and Addition numbers

To multiply and add numbers always first multiply then add

Example 1

$5006 \times 5 + 4970 = 30000$

1) $\begin{array}{r} 5006 \\ \times \underline{\quad 5} \\ 25030 \end{array}$ 2) $\begin{array}{r} 25030 \\ \underline{+\ 4970} \\ 30000 \end{array}$

Example 2

$1960 + 6005 \times 8 = 50000$

1) $\begin{array}{r} 6005 \\ \times \underline{\quad 8} \\ 48040 \end{array}$ 2) $\begin{array}{r} 48040 \\ \underline{+1960} \\ 50000 \end{array}$

Multiplication and Subtraction numbers

To multiply and subtract numbers always first multiply then subtract

Example 3 $506 \times 80 - 30480 = 10000$

1) $\begin{array}{r} 506 \\ \underline{\times\ 80} \\ 40480 \end{array}$ 2) $\begin{array}{r} 40480 \\ \underline{-\ 30480} \\ 10000 \end{array}$

Example 4 $55600 - 608 \times 75 = 10000$

1) $\begin{array}{r} 608 \\ \underline{\times 75} \\ 3040 \\ +\ \underline{42560} \\ 45600 \end{array}$ 2) $\begin{array}{r} 55600 \\ \underline{-\ 45600} \\ 10000 \end{array}$

Summary 5 Chapter 8 Lesson 6

Division and Addition numbers

To divide and add numbers always first divide then add

Example 1

```
        2      1
696 + 1520 ÷ 5 = 1000
      304
1) 5)1520   2)  696
    -15        +304
      20       1000
     -20
```

Example 2

```
         1   2
10020 ÷ 5 + 7996 = 10000
   1) 2004      2) 7996
   5)10020        + 2004
    -10           10000
      020
      -20
```

Division and Subtraction numbers

To divide and subtract numbers always first divide after that subtract

Always do multiplication or division before addition or subtraction

Example 3

```
       1    2
24000 ÷ 80 - 200 = 100
```

a) $24000 \div 80 = 2400 \div 8 = 300$

b) $300 - 200 = 100$

Example 4

```
  2       1
720- 1600 ÷ 80 = 720 – 20 = 700
```

a) $1600 \div 80 = 160 \div 8 = 20$

b) $720 - 20 = 700$

Summary 5 Chapter 8 Lesson 7
Addition with Parentheses
Work always in parentheses first

Example 1

a)$(17 + 19) + 34 = 36 + 34 = 70$

b)$43 + (18 + 19) = 43 + 37 = 80$

Example 2

c) $(25 + 5) \times 8 = 30 \times 8 = 240$

d) $6 \times (125 + 25) = 6 \times 150 = 900$

Addition and Subtraction with Parentheses
Always first, work in parentheses

Examples 3

a) $(700 + 300) - 300= 1000 - 300 = 700$

b) $(700 - 300) + 300 = 400 + 300 = 700$

Example 4

c) $9000 – (500 + 500) = 9000 – 1000 = 8000$

d) $(1200 – 500) – 500 = 700 – 500 = 200$

Summary 5 Chapter 8 Lesson 8
Multiplication and Division with Parentheses
Always first, work in parentheses

Example 1

a) $706 \times (36 \div 4) = 706 \times 9 = 6354$

b) $24028 \div (16 \div 4) = 24028 \div 4 = 6007$

Expressions with Parentheses
Always first, work in parentheses

Examples 2

a) $15 \times (807 – 200) = 15 \times 607 = 9105$

b) $(9000 + 105) \div 15 = 9105 \div 15 = 607$

c) $(17472 - 9000) \div 12 = 8472 \div 12 = 706$

d) $84072 \div (10 + 2) = 84072 \div 12 = 7006$

Summary 5 Chapter 8 Lesson 9
Many-digit numbers
Final Exam 6 A

Show all steps for full credits not just answers.
Each problem is worth 5 points

Evaluate

1) $\begin{array}{r} 700000 \\ \underline{-\ 234567} \end{array}$

2) $\begin{array}{r} 765 \\ \underline{\times 604} \end{array}$

3) $\begin{array}{r} 706 \\ \underline{\times 605} \end{array}$

4) $193884 \div 321$

5) $243714 \div 302$

6) $240240 \div 80 \times 3$

7) $7490 - 830 \times 3$

8) $56049 - 49042 \div 7$

9) $(30 \times 560) \div 8$

10) $289 \div (93 - 76)$

Grading Scale: A =10 points, B=9points, C=8 points, D=7 points, F=0-6 points

Answers Final Exam 6 A

1) 465433 2) 462060 3) 427130 4) 604 5) 807

6) 9009 7) 5000 8) 49043 9) 2100 10) 17

Summary 5 Chapter 8 Lesson 10
Many-digit numbers
Final Exam 6 B

Show all steps for full credits, not just answers.
Each problem is worth 5 points

Evaluate

1) $\begin{array}{r} 600000 \\ \underline{-\ 123456} \end{array}$

2) $\begin{array}{r} 654 \\ \underline{\times 503} \end{array}$

3) $\begin{array}{r} 605 \\ \underline{\times 403} \end{array}$

4) $105630 \div 210$

5) $182508 \div 201$

6) $180180 \div 90 \times 2$

7) $6840 - 920 \times 2$

8) $42036 - 36030 \div 6$

9) $(20 \times 720) \div 9$

10) $256 \div (81 - 65)$

Grading Scale: A =10 points, B=9points, C=8 points, D=7 points, F=0-6 points

Answers Final Exam 6 B

1)	476544	2)	328962	3)	243815	4)	503	5)	908
6)	4004	7)	5000	8)	36031	9)	1600	10)	16

Students!! Attention!!

Adults, help please students 1 – 4 grades to read and understand advice.

When I was in 5th grade I did not like math. I am not always understood math. I was 11 years girl. Next 6th grade I got a very good teacher. She explains very clearly and I understood her very well. I was worked very hard and I was not afraid to ask questions. Final Grade in Algebra in 6th grade I got an A. I said in class that I want to be a teacher of math. But some students said that I am not enough smart to be a math teacher. I said I will be work hard with math. I did it, I got a Master's degree in math and education, I love my students, I was teaching many years, in High School, part-time college math instructor, full-time university math lecturer.

Some people do a simple job, eat, sleep, go to work. People with a bad brain have not so a good job, not so good salary. If you look at TV how someone's dance, you will be not changing your brain. But if you take a dancing class this helps your body and brain ability. The same if you want to use the calculator, but the calculator not always with you. Much better if you learn how you need to do any operations with numbers without a calculator. If you read this math book you will know what you need to do. This teaches you how to improve your memory, attention, logic thinking, how to read your math books every day, our brain works better. You can work 50 minutes with math, 10 minutes break maybe dance. Do not afraid to do mistakes, fixed and move to the next step. Don't listen to anyone, choose what do you like more.

Students need to know if you work with the brain every day, your brain works better. You need a good brain in any job, in any class. When you finish High School, you need to get any profession, you need to go to college and get a good job and a better salary. You need to learn any subject, its work for the brain. Read the book, do the exercises yourself, try your best, and then check your answer at the end of the chapter. Be honest with yourself. You need to believe in yourself. The brain is like muscles. People with good brain lives longer. The professors at my university were over 80 years old, wise, and optimistic. It was a great pleasure to talk to them.